The Cambridge Manuals of Science and
Literature

THE SUN

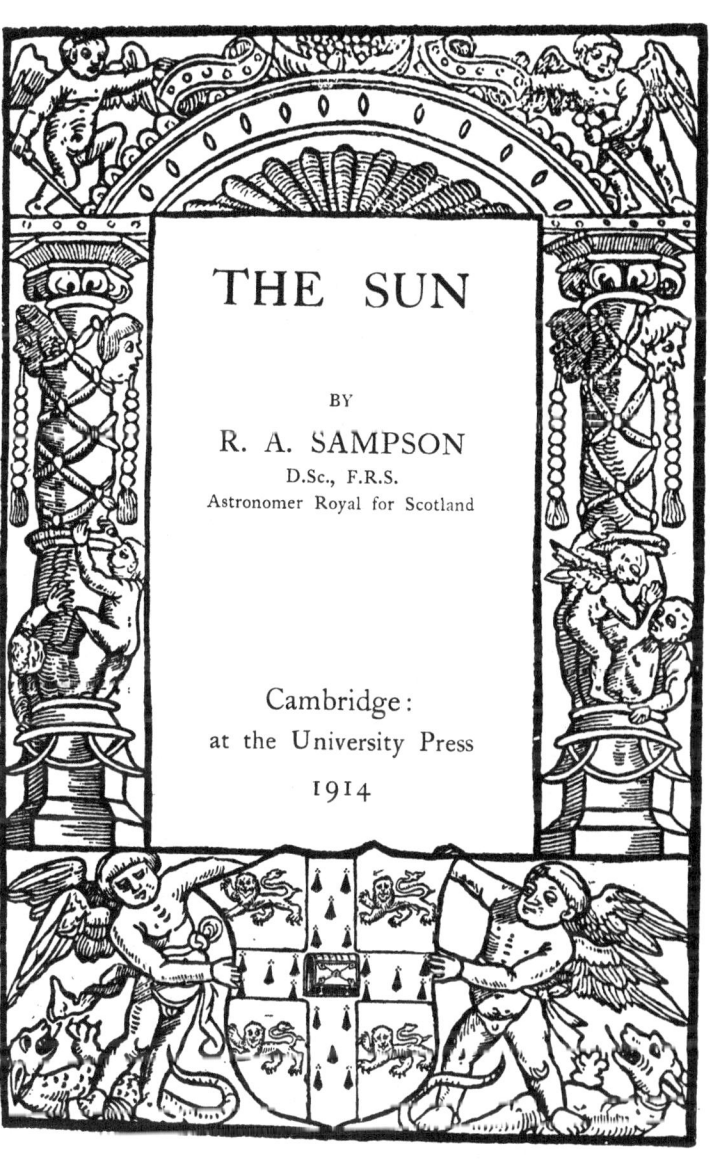

THE SUN

BY

R. A. SAMPSON
D.Sc., F.R.S.
Astronomer Royal for Scotland

Cambridge:
at the University Press
1914

CAMBRIDGE UNIVERSITY PRESS
Cambridge, New York, Melbourne, Madrid, Cape Town,
Singapore, São Paulo, Delhi, Tokyo, Mexico City

Cambridge University Press
The Edinburgh Building, Cambridge CB2 8RU, UK

Published in the United States of America by
Cambridge University Press, New York

www.cambridge.org
Information on this title: www.cambridge.org/9781107401648

© Cambridge University Press 1914

First published 1914
First paperback edition 2011

A catalogue record for this publication is available from the British Library

ISBN 978-1-107-40164-8 Paperback

Cambridge University Press has no responsibility for the persistence or
accuracy of URLs for external or third-party internet websites referred to in
this publication, and does not guarantee that any content on such websites is,
or will remain, accurate or appropriate.

*With the exception of the coat of arms
at the foot, the design on the title page is a
reproduction of one used by the earliest known
Cambridge printer, John Siberch, 1521*

PREFACE

THE aim of this little book is to provide, for the general reader, within the compass permitted, something like a report upon the present position of fact and theory relating to the Sun. I have found this rather difficult to do, and if in the end there remain after all a good many omissions, a good many technicalities, and a good many repetitions of well-known elementary things, I would ask the reader to be generous as well as general, and to remember that the best can only be obtained from a book by taking it somewhat indulgently and upon its own terms.

It is a pleasure to mention here my indebtedness to those who have permitted the reproduction—very inadequate in some cases — of photographs and drawings;—namely to M. Deslandres of Meudon, Prof. G. E. Hale of Mount Wilson, the Royal Society, the Academy of Sciences of St Petersburg, the Royal Astronomical Society, the Carnegie Institution of Washington, the *American Journal of Science*, the *Astrophysical Journal*, and the *Nautical Almanac*. Without their courtesy, slight as the book may be, it would have been impossible to produce it.

R. A. S.

ROYAL OBSERVATORY, EDINBURGH.
November 10, 1913.

CONTENTS

CHAP.		PAGE
	Preface	v
I.	Science and the Sun	1
II.	The Sun's output of heat	12
III.	The Sun as the mechanical centre of the world	30
IV.	The spectroscope	45
V.	Description of the Sun's surface . .	65
VI.	Periodicity of the Sun	82
VII.	Eclipses of the Sun	103
VIII.	The Sun as a star	119
Bibliography		134
Numerical data		137
Index		139

LIST OF ILLUSTRATIONS

FIGURE		PAGE
1.	Langley's Bolograph	16
2.	Orbits of an Earth and a Moon for which five months equal one year . . .	31
3.	Orbits of the Earth, Mars, and Eros . .	38
4.	Fraunhofer's Map of the Solar Spectrum .	48
5.	Fraunhofer's B line, under high dispersion .	55
6.	Pressure displacement in Iron Arc (Duffield). Zeeman Effect on the Iron Spark (King)	62
7.	Langley's Drawing of Sun Spots and details of Solar Surface	68
8.	Direct Photograph of Sun (Meudon), March 21, 1910.	73
9.	Photograph of Sun in Calcium Light (K_3), (Meudon), March 21, 1910 . . .	74
10.	Photograph of Sun in Chromospheric Layer of Hydrogen Light (H_a), (Meudon), March 21, 1910	75
11.	Photograph of Sun in Hydrogen Light (H_a), showing vortices surrounding Sun Spots (Mount Wilson)	77
12.	Photograph of Prominences (H_a) with Sun's disc screened (Mount Wilson), September 20, 1909	78

LIST OF ILLUSTRATIONS

FIGURE		PAGE
13.	Portion of Wolf's diagram of Sun Spot Numbers and Magnetic Variation	85
14.	Schuster's Analysis of Wolf's Sun Spot Numbers for periods between 7 and 18 years	91
15.	Demonstration of the presence of Sun's Rotation Period in Magnetic Records (Chree)	99
16.	Comparison of Variation of Spottedness of Sun with simultaneous variations of ranges of horizontal magnetic force upon the Earth (Chree).	101
17.	Chart from the *Nautical Almanac* showing the limits, and line of central eclipse	104
18.	Hansky's drawings of typical coronal forms, from photographs	112

CHAPTER I

SCIENCE AND THE SUN

Qui addit scientiam, addit et laborem. Eccl. i. 18.

IF we would appreciate scientific results we must understand something of scientific methods; otherwise they are, both one and the other, exposed to the reproach that what they set out to simplify they end in complicating, and when we ask them to explain knowledge which is vague and bring into order what is perplexing, in their attempt to reply they continue to open indefinitely wider and wider vistas for surmise.

Science is another name for knowledge, but it is very different from a mere collection of facts. It is the attempt to understand them and to explain them, as well; to show how the parts spring out of one another, and how far their likenesses are repetitions of one single idea. Briefly it is knowledge, analysed and in order.

Now if we take any simple natural object, such as a flower, let us say a harebell, it possesses a character of its own, an identity, which cannot fail to seize the imagination and which is the primary cause of our wish to know more about it. The slender leaves, slender stalk and bell, the delicate poise, the harmony of the peculiar blue of the bell and peculiar green of the leaf, appeal at once to every eye. But if any lover of flowers, wishing to give shape to this clearly-felt unity, amounting almost to personality, were to seek it in a scientific description of this flower, he would probably be much disappointed. The flower which appeals to him like an emotion, he would find described in words barbarous and clumsy almost beyond parallel, which are heaped upon the little flower till it is crushed beneath them. The mere names of trivial speech, whether we call it harebell or hair-bell, might convey more to him than this heavy-handed work. "There's pansies, that's for thoughts." There, wrapped in a single word, is a whole volume of expression. Why cannot men of science, whose business it is to know what is worth saying, take as their guide some slight but pregnant utterance like that, and fill it out for us?

It is a hopeless request. A moment's reflection will show that, whoever gave that flower its perfect name of pansy or *pensée*, a rapid and incisive stroke of genius like that cannot be touched, added to or analysed, without destroying it. Its force consists not in the meaning it contains, for examined closely it contains no meaning,—nothing but a suggestion, so that each hearer can put his own meaning into it in such a way as pleases him. Now for good and ill that kind of suggestion is what science absolutely rejects. It must make its plans so as to be understood exactly. If we seek a pattern for scientific work we find it not in a phrase from *Hamlet* but in such a book as Euclid's *Elements*, which is so written that one has but to read attentively in order to arrive exactly at those ideas which the author intended to convey, neither more nor less. In some regions the model proves a severe ordinance, but the fertility of science in propagating itself comes from this principle exclusively, by which different workers are enabled to supplement and assist one another and to build upon one another's work as foundation in a way that is unknown in other quarters.

Now before people can be sure that they are upon exactly the same level in comparing their ideas, they must work back to very primitive things. They must dig for foundations down to bedrock. It is the custom in science to draw a line at the metaphysical basis of ideas, though this exclusion is hardly warranted; but apart from that we begin absolutely at the beginning, with space, mass and motion, the chemical elements, and so forth. In other words, when such a theme as the Sun is offered for scientific consideration we do not attempt any united picture of its features in the beginning. We reserve that for the end, though perhaps we never get so far. We begin by picking it into its smallest pieces, and then when we understand these, one by one, we set about the attempt to put them together again. There is no other way possible, and this fact is so familiar and so readily admitted, that the anomalous position in which it puts us hardly strikes us as odd. Yet this process of dissection might well give us pause, for how do we know that in the course of it something may not take flight, like the life from a living thing, that we shall not succeed in putting back into it again?

The plain result is that we do not in the end arrive at any one picture of the Sun, or any other natural object, but at a number of separated and mutually complementary points of view, something after the manner of the ground-plan, elevations and specification of a building. And each of these points of view, if well chosen, will be the starting point of a whole world of its own, of ideas, theories and observations of which the bearing upon the Sun is only an incident. The Sun is indeed in a triple sense the focus of the world; first in the old sense that the Earth revolves about it and regulates its days and seasons by it; next that it is essential to all higher forms of life, its heat ripening the fruits and herbage, but also in the sense that very many of our theories

SCIENCE AND THE SUN

radiate from it, and find in it as in a great physical laboratory their first and most striking application. Thus an adequate theory of motion began with the Sun, in Newton's interpretation of the curvilinear motion of the planets about it, and it is outwards, in direct sequence with the astronomical applications, that still more general methods of dynamics have been reached, culminating in the principle of Least Action, which in a single formula is a summary of the application of dynamics to every physical problem, positing only the original laws of motion. Or to take another illustration, the Sun's perturbations of the Moon's orbit have led in recent times to a wide field, successfully cultivated—the study of the motion of Three Bodies, from a general point of view which is free from the hampering limitations of the actual cases in the heavens, and indeed has probably little or no application there. But the wider theory illuminates the narrower, and the narrower gives a solid starting point for the wider.

Or again, upon the physical side, we have the spectrum. From a certain point of view the Sun only supplies special illustrations of the spectrum, but historically the spectrum originated in sunlight. The dark lines that cross the solar spectrum supplied the origin for all the applications of the spectrum to chemistry. The distribution of energy in the solar spectrum leads to the general theory of the amorphous radiation of the so-called "black body" and back again to an inference as to the Sun's temperature. The same line of advance seems likely to lead to an *experimentum crucis*, the parting of the ways for physics in some of its most elementary conceptions. The Doppler, the Zeeman, and other modifications of the normal spectrum, though not realized first in connection with the solar spectrum, find there some of their most interesting applications. The connection of magnetic fields upon the Earth and upon the Sun is another, though less advanced, question, the solution of which is bound not only to throw light on what we know already, but to bring into existence a train of other unsuspected problems.

Thus it will be seen that the method of dissection and subsequent synthesis, which science has made its own, and which is necessary to its existence, is apt to lead a long way from its own goal and in the end always raises more questions than it settles. It is a necessity to begin by describing a thing in its elementary relations, if people are to be sure that they understand one another. But these relations are general, whereas the problem we start from is particular, and so in the subsequent synthesis, we find ourselves synthesising the general in order to comprehend the particular. To clear up a single known fact, we find a dozen new ones and put them beside it, and each of these may be made a new starting point. Every step forward puts the end of the journey further away. Yet, as a method, it is inevitable unless we are to accept the whole of nature phenomenally, as does an artist.

Convictions like these, I believe, must have occurred to most people and they have a certain hold upon one.

But there is a simpler point of view that is more satisfactory, if one can be content with it. Knowledge in itself is something, and has an attraction for us. It will be sufficient to reflect that the proximate results, the uncompleted applications, are all of great beauty, extent and compelling interest. It is possible that the Whole, if it could be found and stated, would seem arid in comparison, for knowledge is a kaleidoscope that depends upon its repetitions for the pleasure which its patterns can convey.

A few words upon another general matter find their proper place in a first chapter—that is, the part played by speculation in attempting to realize such a phenomenon as the Sun. There is a celebrated phrase of Newton's, towards the end of the second edition of the *Principia*, "hypotheses non fingo," "I make no speculations; for whatever is not deduced from facts must be called a speculation." But this phrase breathes the atmosphere of the formal and Latin *Principia* and is very different from that of the English *Opticks* which concludes with a number of Queries that are nothing more nor less than most penetrating and fertile speculations. In the course of them Newton returns to this question: (Qu. 28)—"the main Business of Natural Philosophy is to argue from Phaenomena without feigning Hypotheses, and to deduce Causes from Effects, till we come to the very first Cause, which is certainly not mechanical. And not only to unfold the Mechanism of the World but chiefly to resolve these and such like Questions....Whence is it that Nature doth nothing in vain, and whence arises all that Order and Beauty that we see in the World?" Now it is clear that such questions will never be resolved by a man whose reach does not exceed his grasp, and speculations are always more interesting than completed demonstrations, for they leave the next step open and invite the newcomer to take part in it. So it is usually the experience of men of science who have something to expound to a general audience that plain information is more welcome if it is tempered by something a little grandiose, some glimpse into another and a better world than that of fact. Rigour and completeness is not only unattractive, it is unsuggestive as well. It is unfortunate that those principles of science that are open to exact demonstration, like that of Least Action mentioned above, or the Principle of Energy, seem to become more jejune the more comprehensive their application is made. If it were the truth that Nature ultimately offered nothing but illustrations of the degradation of energy, it would soon become tedious to verify it. Speculations then must come in, but speculations, of the quality of Newton's Queries, are not to be picked on every bush. These were all intended for proof by demonstration when knowledge should be riper. They followed the course of previous experiment and suggested its continuation. But if speculations are framed not by a Newton, and if they leave a wide scope for construction of details, they are almost sure to be worthless and misleading. Such for instance are the theories which quite a short time ago laid down the circulation of the atmosphere between poles and equator upon the Earth, with ascending and descending currents, to explain the trade winds. These have all gone by the board since we have begun to learn some facts about the actual state of the upper atmosphere. Such details, stepping a good deal beyond knowledge, only tempt men with the semblance of knowledge out of the thorny path of experiment and demonstration which has so slowly yielded them so much. Hence, generally speaking, an attitude of considerable stiffness is observed by scientific men towards speculations. Very few are admitted, and these only to the outer courts, as the hangers-on of lawful members within. Now such a rule if applied to descriptions of the Sun sweeps away a great deal that only a little while ago was regularly used to clothe our naked ignorance. Such are almost all explanations of the Sun spots and their part in solar physics. Such is Herschel's theory of the cool centre and glowing cloak, which I need not apologize for mentioning, for I have myself met people in middle life who had learnt it at school and still believed it.

We should not however make the rule more absolute than necessity demands, for as has been said above it is not an attractive one. Exceptions may be made in certain cases, whether because the subject is too interesting to leave alone, or because the theory in question is associated with some great name and represents the views, even unprovable, of some outstanding man, or chiefly because the lines of research at any particular time run in the direction of working out the amount of truth or otherwise which lies in some particular line of speculation. These exceptions may be dignified with the name of theories rather than speculations, and they should find their place in any living story.

Having said so much about method, a rapid glance may be given to the history of our knowledge of the Sun. It begins with Galileo who, inventing his telescope in 1609, discovered Sun spots in the following year, and among other things has left many beautiful and obviously faithful drawings of great spot groups upon successive days, showing their changes as well as the Sun's rotation. Little or nothing new was added for two centuries, though Sun spots were observed and recorded, here and there, fairly continuously in very many places. Mention may however be made of Wilson's observations (1774) that spots showed a foreshortening at the limb as if they were hollow. W. Herschel was an assiduous observer of the Sun; but his chief contribution to knowledge of it was not the fruit of direct observation, still less his ludicrous theory of its thermal state, but that of assigning its place among the stars by his penetrating determination of its absolute motion in space, first in 1783, confirmed in 1805. A fresh idea was brought into the field by Pouillet, who in 1838 with a clear conception of essentials of the problem, measured the intensity of the Sun's radiation at the Earth's surface. Of all steps made in these comparatively early days the most memorable was the mapping of the dark lines of the solar spectrum by Fraunhofer in 1814. Fraunhofer was not the first to notice these lines, but he was the first to appreciate their importance and to map them in large numbers with proper care. No other avenue of knowledge is in any degree so rich as to the past or so promising for the future. Last I would mention Schwabe's discovery in 1843 of the Sun's periodicity as shown in the growth and decay of its numbers of spots, a periodicity that acquired greatly enhanced importance from its association, discovered by Wolf and Sabine shortly afterwards, with fluctuations of terrestrial magnetism. Each of these heads will form natural divisions of the subject, each representing a separate order of ideas, and leading in every case to very extensive developments in recent years. It will be my purpose to collect, and to some extent to summarize, the present position of these developments, to present them in their right perspective, and in significant relation to one another and to other points of view, though, unavoidably, with many omissions.

CHAPTER II

THE SUN'S OUTPUT OF HEAT

HAVING stated in the last chapter the general aims which we shall pursue in reducing to a scientific nexus the facts relating to the Sun, we begin with the most obvious of all, the Sun's heat, its quantity and its intensity. In reducing to order our ideas upon this matter the first problem that faces us is the part played by the atmosphere of the Earth in the transmitted heat which reaches us. It is obviously a very large part. From one day to the next, with the Sun at the same height in the sky, the amount of heat received by direct radiation upon an instrument shielded from air currents—such as the ordinary black-bulb thermometer, which is exposed within a vacuum—may be utterly different. Even when the sky is apparently clear, the presence or absence of water vapour in the air may change the record entirely. Since these are the circumstances that govern the heat of the Sun as we receive it, they may be more important from a terrestrial point of view than the amount of heat the Sun emits, but as they have nothing to do with the Sun itself, our present business is to eliminate them, and the straightforward method to do so is to make observations, which are effectively simultaneous, about the sea-level, and also at mountain heights which are above the densest portions of the Earth's atmosphere. Prolonged studies are required before it can be said how the amount received is affected by such factors as the altitude of the Sun above the horizon and the time of day, the time of year, as well as the hygrometric state of the atmosphere, and height above the Earth's surface—for all these affect it in ways that can only be determined by experiment. It is found for example that the general absorption is greater in summer than in winter, rising some 20 per cent. from March to August; the afternoon is steadier than the morning, and the most transparent time of day is a little past midday. The great bulk of absorption is due to water vapour.

Research into the Sun's radiations is inseparably connected with the name of S. P. Langley. It was he who invented the instrument, the bolometer, which made an adequate investigation possible, and employed it over a long series of years, first at Alleghany Observatory, of which he was director, and at several other stations, of which the most notable was the summit of Mount Whitney. Mount Whitney is a mountain in the Sierra Nevada range of Southern California, situated in a wild and desert region, where the air is of exceptional dryness. It is also very precipitous, so that it was possible to find stations separated by only a few miles in distance but differing 11,000 feet in altitude. At the upper station— 12,000 feet above the sea-level—about one-third of the atmosphere was below the observer. Here the sky appeared constantly of a deep violet colour, which was continued sharply up to the Sun's limb. The milkiness that is seen all about the disc at lower levels was sensibly absent. Under these circumstances Langley was able to make duplicate observations which permitted him to forecast with some approach to certainty what would be the effect if the observation were made above the atmosphere altogether, and from them it appeared that nearly one-half of what is received from the Sun is

THE SUN'S OUTPUT OF HEAT

absorbed into the atmosphere before the rays reach sea-level.

One point upon which Langley insisted is that the radiations must be explored, not merely as a whole, but for every wave-length, because losses by absorption are different for different wave-lengths. This means that the radiation must be passed through a slit, dispersed into a spectrum, and this spectrum examined with an instrument delicate enough to measure the heat separately transmitted to every point. The bolometer, the instrument which Langley devised for this task, depends in principle on the diminution of the conductivity of a metal to an electric current when the temperature is raised. By setting the two arms of a Wheatstone bridge so that one of them is exposed to the ray and the other is shielded from it, a change in the currents going through these arms may be revealed and measured with a galvanometer. To realize this theory for the indescribably minute quantities of heat that must be investigated naturally requires the extremest delicacy in every part.

Langley was able with its aid to measure the intensity of emission along the whole spectrum throughout the visible portion and four or five times as far beyond in the ultra-red and invisible portion, recognizing by sudden drops in the emission the position of all the chief Fraunhofer lines and discovering

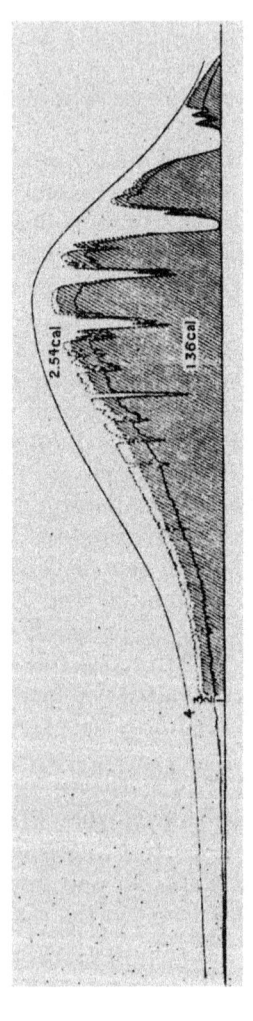

Fig. 1. Langley's Bolograph.

CH. II] THE SUN'S OUTPUT OF HEAT 17

many new absorption lines and bands in the invisible ultra-red. The whole story is most concisely told in the figure, called a bolograph, which embodies his results. The lowest outline represents the actual emissions recorded. The serrations are possibly in part accidental but include indications of the great absorptions which are always present. The steps in interpretation of this record are successively to increase the amounts by allowing for the ascertained absorption of, first, the spectroscope and, second, the siderostat or mirror which reflects the beam of sunlight into a convenient direction; and the final allowance for absorption in the atmosphere gives the smooth curve; its ordinates measure the intensity of emission at each wave-length and its area the total rate of emission of energy. The value of this total will vary very much from one bolograph to another, because the allowance for absorption must refer to an average state of the atmosphere and the actual circumstances may depart from the average very widely and in a way we cannot ascertain. On the whole it seems that Langley overdid his allowances. Langley's investigation, repeated by Abbot, who had worked with him, shows a value decidedly less than Langley gave. It is expressed as the quantity of heat which would be received per minute upon a square centimetre exposed perpendicularly to the Sun's rays above the level of the Earth's atmosphere.

This amount Abbot finds to be enough to raise a gramme of water 1°·95 Centigrade. This refers to the mean distance, for it shows a fluctuation, as it ought to do, as the Earth recedes from or approaches the Sun. It also shows a fluctuation which Abbot believes to belong to the actual emission from the Sun, and of this we shall speak further in a later chapter.

Returning to the figure above, collecting all these emissions from the surface of the vast sphere with centre at the Sun and surface at the Earth, and concentrating them upon the Sun's surface, we should have the radiations as they leave the Sun's body, and they would then be increased in intensity 43,000 times, that is to say, they would suffice to raise a layer of water 9 metres, or 30 feet thick, from the freezing point to the boiling point in a minute. The same amount would melt a layer of 36 feet of ice, or vapourize a layer of 5·4 feet of water. Now it is worth remarking that of all this enormous output of heat the only portion which produces any known effect is the minute fraction intercepted by the discs of the planets, and this is about the hundred and twenty millionth part of the whole. The rest, so far as we can tell, is absolutely lost.

What is the temperature of the Sun? To answer this question we must first connect and coordinate our ideas of temperature and of radiation, which present this essential contrast that while radiation must be looked upon as composed of undulations each of separate wave-length as well as of separate intensity, temperature is featureless in everything except its grade. Temperature tends to equalize itself whether bodies communicate by contact or by radiation, but once an equilibrium of temperature is reached, no disturbance of it can be produced by communication with any different body at this same temperature. Now different bodies certainly differ in their treatment of different wave-lengths of radiation. It must follow then that in the state represented by a steady temperature, the radiations between the bodies must be such that the distribution of energy according to wave-length in them must be wholly independent of the nature of the bodies. That is to say, for each temperature there is a definite distribution of radiation-energy along the spectrum. It remains to ascertain what is the law of this distribution and then we shall be able to infer, in the Sun, for example, from the distribution of energy determined by the bolometer, what temperature must be ascribed to it. This is one way in which we can determine the temperature; we may say that it rests upon a consideration of the *shape* of the bolograph. An independent method depends upon the integrated *area* of the bolograph, namely, we may add together the energy emitted over the whole spectrum and determine terrestrially how such an integrated sum is related to the temperature of the emitting body. To take the latter first. A steady state of radiation corresponding to any given temperature will be established in the interior, say, of a copper globe the surface of which is maintained constantly at that temperature. If a small window be cut in such a globe, and a thermopile placed there, exposed to the radiations, these can be ascertained for the temperature in question. It appears that the energy-density of such radiations is proportional to the fourth power of the temperature, counted from absolute zero; and indeed the same can be established *a priori*. This is the necessary key to the second method. As to the first, the form of the curve in which the energy is distributed along the spectrum for different temperatures has also been determined, partly by theory, partly by experiment, each reinforcing the other to give a sufficiently precise result. In particular it appears that the point of maximum intensity, represented by the summit of the bolograph, shifts towards the violet as the temperature rises, and the temperature is proportional to the wave-frequency at the point of maximum intensity. For the Sun this point occurs about wave-length 4900, near the F line in the yellow-green. The corresponding temperature inferred is just under 6000° absolute, on the Centigrade scale. We reach just the same result if we calculate from the total

THE SUN'S OUTPUT OF HEAT

energy emitted, so that this may be called the temperature of the Sun's surface if we dispense with any further refinements. But there is still the question of the part played by the Sun's own atmosphere. It requires no very careful observation to see that this acts as a cloak, for the disc is visibly darker at the limb than at the centre, and this is due to nothing but absorption within the Sun's atmosphere. Now if the whole mass, atmosphere included, were radiating equally it would appear equally bright. Let us then suppose that there is a nucleus which radiates so, and that it is surrounded by an atmosphere which partly transmits and partly absorbs what reaches it from this hotter nucleus. The progress of this absorption, from the centre of the disc to the limb can be followed—again with the bolograph—for different wave-lengths. For the violet it is more than one-half, for the deep red it is about one-quarter, still decreasing as we pass along to the infra-red. If for the sake of simplicity we suppose the Sun's atmosphere to form a uniform sheet, we find that the radiations corresponding to our temperature of 6000° might equally well have been produced by a nucleus at about 6700° and an atmospheric cloak at an average temperature of about 5500°.

The law of radiation gives another means of estimating the temperature of the Sun. The average temperature at the Earth's surface may be taken as arising from a balance between what it receives from the Sun and what it radiates. If we take this temperature to be 15° C. or say about 290° absolute, then calculate what the Earth would radiate and equate it to the amount received from the Sun, we have an equation for determining the temperature of the Sun. Again we arrive at about 6000°. This argument might seem open to a good many qualifications, as the part played by the Earth's atmosphere in exchanges of heat, the quantity of heat received which is transmitted inwards, and the like. The agreement with other determinations may rather be taken as a verification that the argument is fairly reliable, and from this point of view it is of much interest, for it may be applied equally with certain cautions to the moon and the planets, and will give an indication of the same sort as to their surface temperatures; Mars in this way appears to be about the temperature of freezing mercury, the full moon not far from the boiling point of water, Venus somewhat below this, Neptune not very different from absolute zero.

We cannot form a proper estimate of the mechanical state in the Sun's envelopes without including radiation pressure. It may be shown by theory and verified by experiment that a disc exposed to direct radiation experiences a pressure, which is proportional to the energy-density of the stream and

THE SUN'S OUTPUT OF HEAT

is in fact the same in quantity as if the disc were stopping material particles carrying the same energy, though in its nature the source of the pressure is quite different from this. Upon a small sphere this pressure will vary with the square of the radius. At the same time gravitational attraction will vary with the cube. Hence for any temperature we have only to make the radius small enough and we shall find a point for which the repulsion of radiation is equal to the attraction of gravitation. Within the body of the Sun, when a particle is surrounded on all sides by matter of much the same temperature, these radiation pressures may produce little resultant effect, but in the outer parts there will be a marked and finally a decided balance outwards, which will have the general effect of less density and compression than gravitation alone could produce.

A consideration of the part played by the Sun's atmosphere is of much importance when we study the solar spectrum, which is our chief document; we may ask what layer of the Sun do its absorption lines represent, and generally what do we take to be the mechanism of transmission of energy from the Sun's nucleus, through its atmosphere, to space. A state of atmosphere can be figured, in which the parts are in continual exchange of place, in which the temperature falls away as we recede from the centre, but preserves such a balance that at the lower level where the same matter is more compressed than at a higher level, the relative temperatures are such that the one can expand or contract into the other without loss or gain to the outside. Such a state is called adiabatic equilibrium of temperature. We may start our ideas from it because it may be supposed to be the state into which an atmosphere would settle, more or less exactly, if there were forces continually in operation mingling parts of different layers together; but it can only correspond to strata of the Sun from which not much loss of heat by radiation takes place, because it assumes that matter is transferred from one level to another without such loss. In actuality matter which reaches the higher levels and is therefore less shielded by other strata beyond it, will radiate its heat with less obstruction, and so will fall in temperature below that required for adiabatic equilibrium. It will therefore condense to more than the density due for its level and, being unstable in such a state, will descend, giving place to other portions which in their turn will repeat the same process. In this way it may be seen that the Sun's atmosphere is a mechanism for increasing its effective radiation. But the character of the spectrum forbids us to believe that it is the chief agent, or indeed that anything but a minute fraction of the radiations that reach us come from gases tenuous enough to produce a line spectrum. These thin and relatively cold gases

II] THE SUN'S OUTPUT OF HEAT

that are found in the upper strata are represented by the absorption-lines in the spectrum. These lines are dark only relatively and flash out as bright when the disc is shut off, as during eclipse, they therefore represent some part of the radiations we receive, but the great mass give the continuous spectrum which must originate from matter in a state of density comparable with that of liquids. It can hardly be a true liquid, because every substance has a critical point of temperature beyond which no compression can turn it into the liquid state. But it must be dense. As we recede from this portion and reach regions where pressure is less and temperature lower, absorption-lines will gradually appear, at the lower levels faint and diffuse, at the higher ones strong and sharp, in correspondence with the sharpness of the contrast between the state of the nucleus and that of the layer where absorption takes place.

The process of radiation must have an eventual consequence in the gradual contraction of the whole body of the Sun. Considering only the atmosphere, we have seen how loss of heat by the outer parts will result in a contraction of that part, and each part in turn is exposed to this. Thus the atmosphere is continually contracting upon the nucleus, and drawing out from the nucleus portions charged with heat upon a higher scale than their own, which in turn go through the same process. Thus contraction must of necessity be continually proceeding. And this contraction, paradoxical as the statement may sound, may, and for a period does, raise the temperature of the whole above the level at which it stood before. For a process of contraction, taking place under the gravitation of the parts, is accompanied by a loss of potential energy, since it must be converted into the motion necessary for the contraction. In the compressed state of the Sun, this motion can result only in collisions among the particles and will at once degenerate into heat. This process will generally and approximately keep pace with the loss by radiation which sets it in action; but always a kinetic process, produced by some disturbance of equilibrium, outruns the equilibrium position which it aims at restoring; and if, as by conversion of motion through collisions into heat, it then produces an irreversible state, its effects will gradually accumulate. Thus as the Sun contracts by radiating heat, more heat is derived from the contraction than would replace the heat lost by radiation. In other words, a sun gets hotter as it contracts, at any rate for a period; at last there must be a limit to the process, as there must be supposed to be a limit to the point to which contraction can proceed. When this limit is reached the radiation, which still proceeds, will be uncompensated, and the whole body will gradually grow colder. At what stage the Sun may be in this

THE SUN'S OUTPUT OF HEAT

process must remain conjectural; its density is about one-fourth part of that of the Earth; hence it seems still far from the limit of its contraction, and is therefore probably still growing hotter.

The amount of contraction that would be necessary just to replace the loss of heat by radiation can be calculated, at any rate if we make some simple assumption as to the distribution of density within the Sun's body. Say that the density is uniform, and we find the amount is no more than an inch a year. Such an amount is utterly insensible, and the Sun's radiation might be made good by contraction throughout any range of historical time without any chance of our being able to detect it. Yet if we push the idea to extreme consequences the rate is too rapid to fit in well with other well established indications of science. For looking into the past, if the contraction was always just enough to replace a loss of heat at the present rate, it follows that at an assignable epoch in the past the process must have begun, with the matter of the Sun all in a state as diffuse as that of a nebula. When the calculation is made the date that emerges is 17 million years ago, an immense space of time, but not nearly great enough. The Earth cannot be older than the Sun, but unless all indications of geology are astray, the history of the Earth presents almost endless vistas of slow changes that could not possibly be compressed within 17 million years. The matter is puzzling. Of course it may be that the contraction of a nebula is at first very slow, and its rate of loss by radiation, small; but even if we confine ourselves to that small portion of the history of the Sun when his condition was not very different from the present, and so the rate of radiation and of contraction much what it is to-day, we have to face the fact that even life has existed upon the Earth for too long a time to allow us so to get out of the difficulty. For we cannot suppose life to have existed with a Sun utterly different from the one we know. Many suggestions have been made to assist the maintenance of the Sun's heat. One of these is the fall of meteors, which would carry with them a contribution of energy. It is true that meteoric dust, in sensible amount, is continually falling upon the Earth, and much more must fall upon the Sun. But most meteors, like most comets, would not fall into the Sun, but would circulate about it; they would in bulk, and in the accumulations of millions of years, be of large mass, and we could hardly fail to trace their effects in disturbance of the planets' motions. Besides, this continual accretion would imply a very different sun from the present one in the remote days to which we are seeking to apply our theory.

It is perhaps hardly necessary to say that maintenance by combustion is out of the question. The

THE SUN'S OUTPUT OF HEAT

heat of combustion is energy set free in the rearrangement of molecules of two substances when they form a compound of lower intrinsic energy than the sum of their separate energies. It cannot be said that there are no chemical compounds upon the Sun, since titanium oxide and manganese hydride and cyanogen have been detected; but the temperature of the Sun is too high to allow of the formation of nearly all compounds that we know, and there is certainly no evidence that this has been a substantial source in the past. But a similar idea looked at from a different point seems more hopeful. The discoveries of radio-activity prove that under certain circumstances it is not impossible to draw upon the energy within the atoms of the elements. For all we know such energy may be large.

It is quite possible that in the course of the development of the Sun from a primitive state, the chemical elements such as we now know them in the Sun, have undergone development also, by rearrangements of the electrons that build up their atoms, and that this has formed an important source of supply of the Sun's heat in the past. If this were so its radiation may have been maintained for an indefinite period in the past, at its present rate, without any great shrinkage of the body.

CHAPTER III

THE SUN AS THE MECHANICAL CENTRE OF THE WORLD

THE Sun dominates the world with its light and heat, but in an even subtler way he dominates it with his mass. At a distance of no more than twenty diameters from the Earth's surface, a body, left to itself in space, between the Earth and the Sun, would fall to the Sun and not to the Earth, though the former is nearly six hundred times further away. In this sense the Moon may be said to be governed more by the Sun than by the Earth for its distance from the Earth exceeds by fifty per cent. this limit. Indeed the Moon might quite correctly be described as one of the planets. As it accompanies the Earth in its annual course around the Sun, in some configurations the Earth and Sun pull upon it in the same direction, and in other configurations they pull in opposite directions, and in the latter case the Sun always predominates over the Earth, nearly in the ratio of two to one. From a heliocentric point of view, Earth and Moon each describe an annual orbit, approximately elliptical but with twelve or thirteen flattenings and the same number of rounded corners, marking the places where the one is within or without the other. Because the Earth is rather more

III] SUN MECHANICAL CENTRE OF WORLD 31

than eighty times as massive as the Moon, the orbit of the Earth is eighty times less flattened or rounded than that of the Moon. A flattening of one orbit

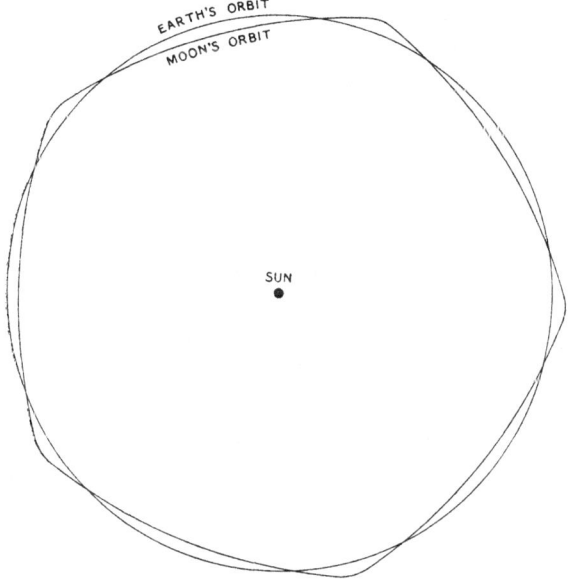

Fig. 2. Orbits of an Earth and a Moon for which five months equal one year.

corresponds to a rounding of the other orbit, so that while the Moon is falling towards the Sun and therefore away from an Earth supposed stationary, the

Earth at the same moment is falling towards the Sun faster than the Moon. Hence the relative effect is produced of the Moon still falling towards the Earth. The orbits are thus so intertwined that from the Earth the Moon seems to circulate round the Earth, while from the Moon the Earth would seem to circulate round the Moon. The disproportion of the numbers is too great to allow of a figure which will show these facts correctly, but they will be understood from the diagram given which is drawn for a moon which made five months to a year in place of twelve and a half.

Even at the Earth's surface, where, of course, the direct attraction of the Earth upon bodies greatly preponderates over that of the Sun, the latter is still very considerable. It amounts to over 21 ounces in the ton. At midday at a place with the Sun immediately overhead the pull of a mass of a ton towards the Earth is diminished by this amount; at midnight it is increased by the same quantity. So large a variation could under most circumstances easily be made sensible by some laboratory test, but in this case such a course is not possible, because the whole of this pull is exactly spent in making that particular ton describe its annual orbit round the Sun, and since we cannot prevent it from doing so, there is no phenomenon, other than this circulation round the Sun, left to investigate.

III] SUN MECHANICAL CENTRE OF WORLD

It is not very difficult to compare the mass of the Sun with that of the Earth, laying aside refinements. The Earth's attraction for the Moon gives the link necessary for the comparison. It can be shown that for any satellite, round any central body, a measure of the period of revolution is the distance, multiplied by the square root of the same, and divided by the square root of the mass. Now the Earth is about 400 times more distant from the Sun than the Moon is from the Earth, and $400 \times \sqrt{400} = 8000$. Again the period of the Earth about the Sun is about thirteen and a half times that of the Moon about the Earth. Therefore the mass of the Sun is about $(8000/13\frac{1}{2})^2$, or 350,000 times that of the Earth. The exact value, when all allowances are made that may qualify this argument, is 332,000, or, in comparison with the Earth and Moon together, 328,000.

To give no greater mass than this within its actual volume, the average density of the Sun must be decidedly below that of the Earth. For the radius of the Sun as seen from the Earth is $961''$ while that of the Earth as seen from the Sun is $8''·80$, and the volumes are as the cubes of these quantities, namely 109^3 to 1, or in the ratio of 1,300,000 to 1. Hence upon the average the Earth is about 4·2 times as dense as the Sun. Now the mean density of the Earth can be found by comparing its attraction, that is to say, gravity, with the very minute attraction for one another of balls of lead, as was first done by Cavendish, and it appears to be about 5·5 that of water. Hence the average density of the Sun is about 1·4 times that of water. It must be clearly recognized that this is only an average. The outer coat of the Sun is gaseous and tenuous enough to yield sharp lines as a spectrum. What proportion of the Sun's body is thus constituted is largely a matter of theory or even speculation and naturally the lower the density of the outer parts, the greater must be that within, in order to make up the average. It seems probable however that this density may be fairly representative of the great bulk of the body. We must remember in the first place how great will be the attraction of the Sun upon its own atmosphere. The intensity of gravity at the Sun's surface is twenty-seven times as great as it is at the Earth's surface. A falling body would fall over 440 feet in the first second. On the Earth an atmosphere which seems to be negligible above a height of about 30 miles reaches at the Earth's surface a density about one thousandth of that of water or one six-thousandth of the density of the globe. Arguing from the relative intensity of gravity, the corresponding figures for the Sun would give, say, one quarter of that of water or one-fifth of that of the globe. It seems probable that relatively quite a short way from the visible outer margin the Sun approaches its mean

III] SUN MECHANICAL CENTRE OF WORLD

density though this conclusion is obscured by an unknown allowance for the diffusing effect of radiation-pressure. Moreover, as explained in the last chapter, it appears that the outer parts, cooling and condensing will be continually exchanging their position with matter which lies deeper down. And this is a phenomenon which must extend in some degree throughout the whole body since no part of it can be truly solid at the temperatures we know to subsist. There will therefore be a continual process of stirring the matter of which the Sun is constituted and it seems unlikely that any great inequalities of density should remain. The pressure of course must increase greatly inwards but as far as an effect in increasing density is concerned it will be offset to a greater or less degree, how much we cannot say, by increased temperature.

The Sun's semi-diameter is always measured in angle and the statement that this angle is of the average value of 960″ is merely an expression of the ratio of the true linear semi-diameter to the Sun's distance. For example it will vary with the distance, being greater by 16″ in winter and less by the same amount in summer. To make a comparison of the actual dimensions of the Sun with those of the Earth, as has been done above, we want to know as well what angle would measure the semi-diameter of the Earth as seen from the Sun—which is technically known as the Sun's parallax. The ratio of the angular value of the Sun's semi-diameter to the parallax is a constant and is equal to the ratio of the Sun's radius to that of the Earth, viz. 109 : 1, as stated above. The angular semi-diameter of the Sun is comparatively easy to determine, the chief difficulty being that, seen through the lower atmosphere, under ordinary circumstances, the limb appears to "boil" and renders exact measures a little uncertain. Still by collecting and discussing critically a great number of such measures a high degree of certainty has been reached. It is then found that the disc is sensibly a true circle, without any of the equatorial protuberance that is found in the Earth, Jupiter and Saturn. Also that there is no sensible fluctuation in the measures—apart from what is due merely to the varying distance of the observer—such as one might conceivably suppose to accompany the sun spot period.

To find the Sun's parallax is a very different affair, that can only be approached in a circuitous fashion. But the importance of the determination matches its difficulty, for if we know the Earth's angular semi-diameter as it appears from the Sun, we can infer the distance of the Sun, and we have the link that connects terrestrial and celestial dimensions, not only of the solar system but also of the stellar universe. The process of determination, though intricate, is worth notice, as illustrative of the methods of astronomy.

III] SUN MECHANICAL CENTRE OF WORLD 37

Naturally we cannot take our station at the Sun and measure from there the angle which the Earth's semi-diameter subtends. We must instead infer this, the vertical, angle of a long triangle by measuring the two base angles at the Earth, and even this cannot be done directly, for two distant points on the Earth are out of sight from one another. Apart from this the observations could with difficulty be made simultaneously, and if they were not, the Earth's rotation in the interval between them would carry the observers into places that broke up the triangle originally contemplated. Besides, the Sun itself is ill-suited for observations of minute changes of position, since the stars, which are the fixed reference points, are not visible along with it in the sky. These and similar reasons made the question of the Sun's true distance a very ill-solved problem up to quite recent days. Without entering into all the steps it may be indicated how the accumulation of knowledge in other directions and the coordination which is so characteristic of the astronomy of to-day has gradually overcome these difficulties.

First, it is unnecessary to observe the Sun directly. Any member of the planetary system will serve as well or better, provided we can compare its distance with that of the Sun, and Kepler's laws, or more accurately the law of gravitation, permit us to express the ratio of the distances of the various planets when we know their angular motions, and these are easy to observe with all necessary accuracy. Mars is not an unsuitable planet for this purpose, but certain of the minor planets are much better and much the best of these is the 433rd in order of discovery, the planet

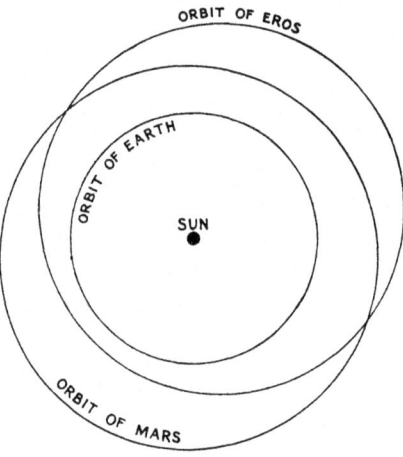

Fig. 3. Orbits of the Earth, Mars, and Eros.

Eros, which comes upon favourable occasions within less than one-sixth of the distance of the Sun from the Earth and therefore increases sixfold the small angle which we seek to measure. It is moreover in appearance like a very faint star and so is suitable for measurement alongside the true stars. Next,

III] SUN MECHANICAL CENTRE OF WORLD

simultaneous observations may be dispensed with if we know the motion well enough to allow for it in the interval between the observations, and thus the same observer at the same spot upon the Earth may make observations from both ends of the base line, letting the rotation of the Earth carry him from one end of it to the other between evening and morning, and this is a great advantage, for besides freeing the results from differences due to personal habits—what astronomers call the "personal equation"—the fixed stars employed for reference may be the same from evening to the following morning. Thus many difficulties of coordination are avoided. But to get the fullest possible use of an occasion such as the near approach of Eros, which will not return for thirty years, in spite of difficulties, the coordination of all possible material must be faced in all its complexity. Photographs of the field containing the planet were obtained on all possible occasions by a dozen observatories situated all about the world, during the whole time while Eros was near enough to render observations profitable, namely some four months. During this time the planet passed over an arc of 50° upon the sky. It might be imagined that all the stars it would pass in its course would be thoroughly known, but that is not the case. They are for the most part insignificant stars which there was no reason to observe closely. Moreover, when we come to points of high precision, the place assigned to a star by different observers with different instruments, in spite of all care, is not the same. It is a matter of great difficulty to decide what is to be done with such differences. Nothing will remove the discrepancies, small though these be, and since some single result must be adopted a decision must be made of the best way to smooth them over. This is the chief and essential difficulty of bringing the whole of the different observations into unison; but there are many others, which, though matters of detail, are serious enough. The theoretical place of the planet must be calculated with extreme accuracy, so that it may be compared with the observed place. And so forth. In the end when all the observations have been collected the uncertainty of the result is indeed diminished, but only to about two-thirds of what can be got with incomparably less toil from the discussion of night-and-morning observations of a single observatory. The concluded result is that the Earth's radius at the equator subtends at the Sun's centre an angle $8''\cdot807$. Of this the last figure is uncertain. Different methods of discussion of the material diminish it by a few units. The measure of uncertainty as represented by the so-called probable error is $\pm 0''\cdot0027$. This gives the Sun at a mean distance of 92,830,000 miles, with an uncertainty of 30,000 miles one way or the

III] SUN MECHANICAL CENTRE OF WORLD 41

other, an amount that must be considered truly minute, since a mile, more or less, in the Earth's radius would produce an equal change.

The Sun's distance when found gives the scale of the solar system, and thence the scale of the stellar system as far as we can probe the latter. Now we have seen that the Sun's distance depends upon the Earth's radius, and it is worth remembering that all we know of the Earth's radius is built upon the actual measurement of a few base lines, a few miles long, upon the surface of the Earth.

What is the nature of the tie that binds the Earth to the Sun across this immense void? It is altogether mysterious. Newton who demonstrated its existence left on record his view, that no one competent to think upon the subject could imagine that it was not somehow exerted through and throughout an intervening medium. But we can see no trace of the medium about it. Light, the swiftest moving thing we know, shows trace of the medium through which it is propagated in the time it takes to pass from point to point. We can trace nothing of the kind in gravitation. If there is any gradual propagation, its velocity must be fabulously great. Again there is no such thing as a screen for gravitation. If three bodies are in a line with one another, the outer ones appear to act upon one another as though the middle one were not there; at the same time they act upon the middle one as if it alone were in view from them. The sole statement of the gravity-relationship makes each body, and each particle imbedded in the body, act upon every particle of every other body, simply in virtue of their common existence in space, and this statement, owing to its simplicity, is open to stringent test, for its consequences can be calculated. They have been calculated in numberless cases, for on this sole formula astrodynamics is built, and all our predictions of the places of the heavenly bodies are dependent upon it. Every comparison of agreement of an observed with a predicted place is a test of it and a test which it abundantly satisfies. There are, it is true, two or three obscure and difficult corners where we do not seem able to account for the facts by mere gravitation according to the inverse square. Thus if the Moon's place is calculated with all possible care and compared with a long series of the best observations, it seems that there remains an unaccounted-for fluctuation taking place in the course of two or three hundred years which carries it a second of time to and fro across its computed place. Every phenomenon in the heavens must be put through a drag-net before we find this remaining, almost ridiculous, exception, which we could explain hypothetically in a number of ways. It is more than anything else a testimony to the certainty and completeness of gravitational astronomy that we allow ourselves to

III] SUN MECHANICAL CENTRE OF WORLD 43

speak with any decision upon a basis apparently so slender. It is small wonder that astronomers, on the practical side, are content to treat the law of gravitation as absolute. But upon the theoretical side, as enquirers into the nature of the universe, they are still at a stand in face of Newton's unqualified dictum that no one with a competent faculty of philosophising would hold that gravity is immediately exercised, that is without the intervention of the medium.

In a similar category with gravity we must class another phenomenon which also manifests itself directly from the motion of the Earth about the Sun, and which presents unexplained features, deeply planted in the nature of things. This is the aberration of light. Roughly speaking it is a phenomenon connected with the fact that the speed of light is not infinitely great compared with the speed of the Earth in its orbit. When one is in a railway train, owing to the motion of the train, any shower appears to drive more or less from the quarter towards which the train is travelling. In somewhat the same way it is observed that the light of all the stars reaches us displaced towards the "Earth's way"; a point towards which the Earth is travelling; it is comparatively easy to verify this, because the Earth's way is now in one direction and now in another, and so all the stars must be found, in sympathy, appearing to circulate annually about their mean positions.

This verifies itself, but of course the analogy with drops of rain is not so much rough as altogether misleading. As a theory it cannot be said to do more than "save appearances." The motion conveyed by the drop is isolated and affects in no way the medium through which it passes. But the motion conveyed in waves of light is handed on from one stratum of the propagating medium to the next; and it would scatter at once if it were not restrained to a definite passage by the corresponding motion propagated in the neighbouring parts of the wave-front. It must be regarded as an affair of the whole wave-front, and not as belonging to a geometrical line drawn from the Earth to the star. If the medium is disturbed the wave-front will be disturbed. Now as the Earth moves through the medium, does it produce no disturbance at all of the wave-front? It seems hardly credible, though aberration is not the only phenomenon that seems to assert it. There is another point. We know very well the velocity of light, and we can calculate the velocity of the Earth in its orbit, from the value of the Sun's distance taken as determined above, and from these we can calculate the aberrational displacement we should expect in a star. It does not agree well with what is observed, and what seems well established—though it should be added that these observations are encumbered by some disturbances that are difficult to

III] SUN MECHANICAL CENTRE OF WORLD 45

allow for. To make an agreement we should require a parallax of $8''\cdot 78$ and this is a departure from the accepted value of $8''\cdot 80$ which can hardly be tolerated. Whether these two puzzles are destined to supply one another's solution, or not, must for the moment remain unanswered.

CHAPTER IV

THE SPECTROSCOPE

LITTLE or no progress could be made in understanding the Sun without the spectroscope. Whether we deal with temperature and emission of heat, determination of the elements of which its atmosphere is composed, distribution of these elements, their movements in eddies, jets and drifts, and even the exact determination of the rotation of the whole body, the spectroscope in one or other of its forms is either our sole, or our best, means of knowledge. Its function is to unravel and present, one by one, the elements which combine to make up the Sun's general radiation. It may be illustrated by considering the parallel case of analysing a mixed sound. If, in a concert hall, a phonograph receiver is operated, it will record all that reaches it, and the disc containing the record can be made to reproduce in turn the sound in which preliminary taps of the conductor's baton, the different instruments, the human voice, the applause of the audience, may be easily distinguished. But the record of each composite sound is single. If we imagine a phonograph which, in place of an indentation upon a disc, produced a record like a written page of music, in which every note with its proper strength was separately given, we have the analogy of the spectroscope. We are, as it were, people deaf from birth, who find themselves in a library of music, with the problem of discovering what it means. But with this difference, that most spectrograms are more intricate than any music that was ever written. It is less surprising that progress is slow than that so much has already been made.

The form of spectroscope usually employed with the Sun is a so-called "grating." In its original form, invented and made by Fraunhofer, who first mapped the lines in the solar spectrum it was an actual grating, formed with the finest wires stretched parallel to one another at the narrowest possible intervals across a frame. When the front of a wave of light meets such an obstacle, it ceases to progress in one piece as before; it is broken up, and in place of a single wave with its front all in the same phase and proceeding outwards radially from its original source, we have, side by side, a large number of sources, which are all in the same phase, and which send out light in all other directions besides that of the original propagation. Now in oblique directions, the point where the light is received will be at different distances from the different sources and hence the waves from these will reach it in different phases. If the grating is very regularly spaced and the point some little distance away these differences of distance will be constant. Consider then that colour of light for which this constant difference for a particular oblique direction is just one wave-length. For this colour all the waves which reach the point will be in the same phase and will reinforce one another. This colour will appear with full intensity in this particular direction; but if we vary the wave-length ever so little, the phases of the different partial waves will be sufficiently different to render the net result null. At an adjacent point the case will be reversed; we shall find waves of the second length at full strength while those of the first length are annulled. Thus the grating will effect a resolution of the light into its different constituent wave-lengths. The refinement with which it does so is limited only by the mechanical difficulty of constructing a fine grating. Those constructed by Rowland have never been surpassed. To give an illustration from an actual example—a fine one, though by no means exceptional in its excellence—we may describe the Rowland grating in use at the Royal Observatory,

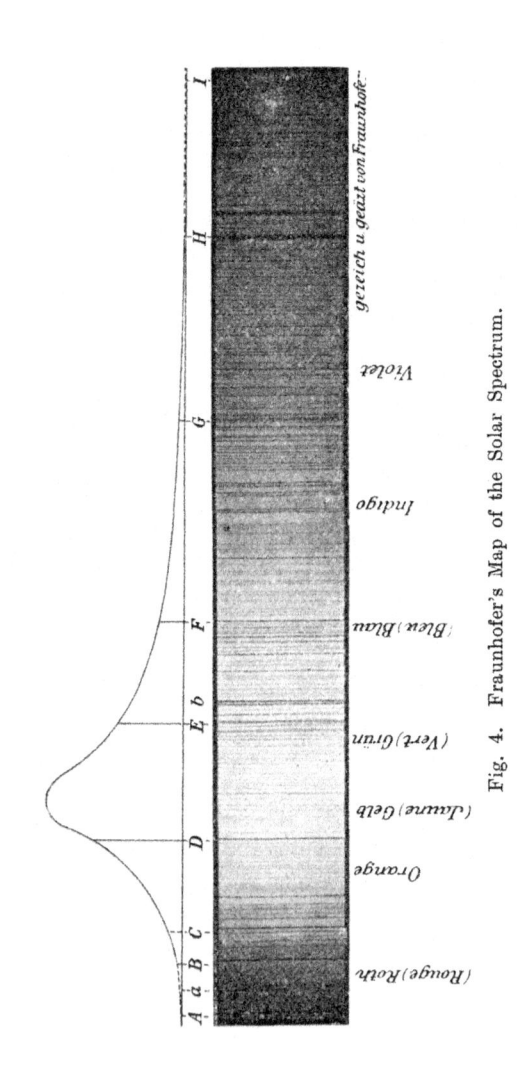

Fig. 4. Fraunhofer's Map of the Solar Spectrum.

CH. IV] THE SPECTROSCOPE 49

Edinburgh. A piece of speculum metal is polished plane and ruled for 5½ inches by parallel lines 4 inches long, 14,438 lines to every inch. With our collimators this gives a visible spectrum between 2 and 3 feet long, if we keep to the first order in which the successive differences equal a single wave-length. If we pass to the second and third orders where the differences are respectively two and three wave-lengths and which are therefore seen more obliquely, the lengths are 4 feet and 7 feet respectively. This spectrum consists of side-by-side images of the slit through which the light passes, and this slit may be, say, one-thousandth of an inch in breadth, so that the third-order spectrum would separate and record independently some seventy-five thousand shades of distinction in the original light, and the relative positions of the lines recorded may be measured with about ten times this precision.

The solar spectrum would be of comparatively little use, were it not for the gaps in it. These gaps or dark lines, noticed first by Wollaston and discovered independently, mapped and named by Fraunhofer, after whom they are called, offer definite measuring marks, upwards of 20,000 of which, in the visible spectrum and its continuations into the ultra-violet and infra-red, have been determined.

The ordinary continuous spectrum gives nothing to lay hold of, and Fraunhofer was seeking to remedy this and improve the very vague determinations of refractive index, by narrowing the slit so as to produce monochromatic lights, when he discovered these natural standards, now perhaps the best known, most precise, and most convenient measuring marks in nature. It is necessary to understand exactly what these dark lines represent. From a general standpoint, they are an illustration of the phenomenon of resonance or sympathetic vibration. Let a system A be tuned to emit a certain note. Then if it is exposed to mixed vibrations, those which will affect it preeminently and the energy of which it will absorb, are just those which it is tuned to emit. The mixed wave will pass on, poorer in respect to the species of wave absorbed by A; or rather in respect to a portion of this, for A will itself emit a portion of what it has absorbed. If, before the passage of the wave, A was emitting radiations, it will emit them more strongly after, but it may be, and usually will be, in negligible amount, compared to the intensity of wave from which it derives them, and which is seen in full force for all the vibrations to which A is not exactly tuned. To apply this to the spectrum, we must regard the chemical elements as the tuned resonators. If any one of these elements, say iron or nickel, is excited to luminous emission by employing it as a pole for the spark of an induction coil, or by placing a portion of a salt of it in the pole of an electric arc, the light emitted consists of a number, sometimes a very great number, of sharp bright lines, representing the periods of vibration with which the element in question is in tune.

In the case of some elements, as for example sodium, the characteristic lines may be generated by simply touching the wick of a spirit lamp with common salt. In such a case the emission is comparatively feeble in comparison, let us say, with the flame of gas, and if the sodium light is placed so that the light from the gas passes through it, the sodium lines will be relatively dark in comparison with the rest of the spectrum, though in fact they are brighter than they were before. The continuous spectrum is yielded by bodies which are sufficiently dense, whether solid, liquid or gaseous. The inference then from the spectrum of the Sun is that it consists of a central mass in a compressed gaseous state—for the temperature is too high to permit it to be liquid—surrounded by an atmosphere the constitution of which we can assign if we can identify the Fraunhofer lines with the emission lines of terrestrial spectra.

The first element identified on these principles was sodium, whose presence in the Sun was taught by Stokes. The whole subject, more fully investigated by Kirchhoff, has led to the recognition with all grades from certainty to mere presumption, of the presence of 45 or 46 terrestrial elements in the Sun's atmosphere. The following is the list, arranged approximately in the order following the number of identified lines.

Iron	Neodymium	Aluminium	Bismuth (?)
Nickel	Yttrium	Cadmium	Tellurium
Titanium	Lanthanum	Rhodium	Indium
Manganese	Niobium	Erbium	Oxygen
Chromium	Molybdenum	Zinc	Tungsten
Cobalt	Palladium	Copper	Mercury (?)
Carbon	Magnesium	Silver	———
Vanadium	Sodium	Germanium	Helium
Zirconium	Silicon	Glucinum	Ytterbium
Cerium	Hydrogen	Tin	Europium
Calcium	Strontium	Lead	Radium (?)
Scandium	Barium	Potassium	

The four last belong to the chromosphere or upper atmosphere.

It must not be taken that the order in this table represents a relative degree of certainty or uncertainty in the identification. No element is more certainly present in the Sun than hydrogen, which comes well down in the middle of the list. It comes low because its spectrum contains few lines—while iron, which heads the list, has a spectrum containing an enormous number. Altogether these identifications represent rather more than one quarter of the lines of the solar spectrum, and of these some 2000 lines are iron lines.

The identification is far from being, in all cases, a straightforward matter. Among the hosts of lines belonging to different elements there are many approximate coincidences. Thus for example nearly all the lines which are now by some attributed to radium were, before its discovery, with as much plausibility ascribed to other elements. And on the converse side because a coincidence of two lines in two spectra is not exact it cannot be certainly inferred that they do not originate from the same element, because there are numerous causes—dealt with below —which produce more or less considerable shifts in the lines of a spectrum, and some of these affect different lines of the same spectrum in different degrees. There is moreover the question whether an absorption line is solar or terrestrial in origin; for naturally the Earth's atmosphere, through which the light passes, leaves its mark upon the spectrum, and this, for example, for long obscured the question of the presence of oxygen. Again nitrogen does not appear in the list, as directly identified, but it must be present because cyanogen, which is a compound of nitrogen, has been recognized. Thus it may be less easy to infer the absence of an element than it is to infer its presence.

The spectrum of an element is dependent upon the mode of production. As a rule the spectra generated by the arc and by the spark are quite different. It seems more likely than not that in the circumstances of the Sun's photosphere, which we can hardly hope to imitate, elements like nitrogen, chlorine and others may be represented by spectra which we have no present means of recognizing.

An absorption line will assert itself strongly in proportion to the contrast, chiefly the temperature contrast, between the state of the absorbing body and that of the emitting body. Hydrogen reaches to all but the highest layers of the Sun's atmosphere, where its temperature will be much below that of the lower photosphere. Hence the hydrogen lines are among the strongest lines in the solar spectrum. But if we take elements like silver, lead, or mercury, of high atomic weight, which we should expect to find in the lower strata, where they would show contrast less marked between their circumstances and those which give rise to the continuous spectrum, we should expect in consequence that their absorption lines would be weak or doubtful; and this is borne out by the fact. But the argument, direct as it is, is not without difficulty in its application. Thus helium, long before it was found upon the Earth was recognized, and named, from the presence of a strong bright line in the photographs of the upper solar atmosphere taken with an eclipsed Sun—for an eclipse of the central disc allows the emission from the atmosphere to show. At the same time eighteen or

IV] THE SPECTROSCOPE 55

twenty hydrogen lines, in their rhythmic series, appear. From analogy with hydrogen we should expect a strong absorption line of helium in the ordinary spectrum. No such thing is found. It is with difficulty that any absorption at the indicated spot can be recognized at all.

To separate the contribution of the Earth's atmosphere to the absorption spectrum from the lines which are solar in origin is essential. It was early recognized that two great absorption bands in the red, Fraunhofer's A and B, the so-called rain bands, became more pronounced when the atmosphere was charged with water vapour. One of these bands, B, is shown from a photograph taken at Edinburgh, upon an enlarged scale. It will be remarked how the undistinguished shading of Fraunhofer's scale

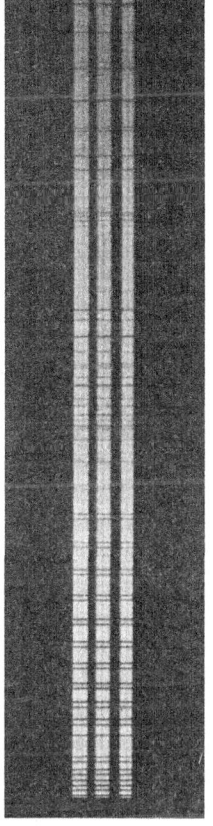

Fig. 5. Fraunhofer's B line, under high dispersion.

when examined with greater power spreads out into a rhythmical arrangement of pairs of lines, which have been successfully analysed arithmetically by Deslandres and Higgs, evidently containing some physical harmony, but we have not yet succeeded in saying precisely what. These lines are merely terrestrial. They can be recognized as generally sharper than the solar lines, though this feature is better shown in other examples where the absorption is not so heavy. This difference of sharpness is one means, often a ready and sufficient means, of recognizing atmospheric lines. Another method is to take the spectrum under similar circumstances when the Sun is high and when it is low. For low Sun the rays pass through a greater thickness of the Earth's atmosphere, and hence telluric lines become more prominent. A third method is to compare spectra taken about the sea-level with that from the top of some mountain, where the station is above the greater part of the atmosphere. With this view Piazzi Smyth observed from the Peak of Teneriffe in 1856 and Janssen from the summit of Mont Blanc in 1888. One or other of these methods has served for all except the most difficult cases, and for example has demonstrated the telluric origin of several oxygen lines which seemed at first to prove the presence of oxygen in the Sun. It became difficult to say whether there demonstrably was or was not any free oxygen

IV] THE SPECTROSCOPE 57

in the Sun, and it may be of interest to describe how the question was settled.

Eliminating one case after another attention was concentrated upon some lines which are barely visible in the deep infra-red. They are too difficult for any of the tests above. In fact it is doubtful if they have ever been seen, but they are recorded upon Higgs's beautiful photographic map of the solar spectrum. Now the spectrum of oxygen, produced in the vacuum tube, runs in a series of characteristic triplets. In 1896, Runge and Paschen, looking in the infra-red for an oxygen line which had been detected by Piazzi Smyth in 1883, remarked that it was not a single line but one of these triplets, and the triplet was recognizable in the proper place in Higgs's map. Because the circumstances in the vacuum tube where alone they are produced have, seemingly, little resemblance to those of the atmosphere, there was a strong probability that the lines were really solar. But it remained merely a presumption, though a strong one, until quite recently, when at length they were proved to be indubitably solar because simultaneous photographs of them, from the east and west limbs of the Sun, showed the relative displacement, known as the Doppler displacement, shown by advancing or receding sources of emission.

The Doppler displacement, to which we now turn, must be treated in some detail. The grating treats the different waves not according to their periods, which are invariable, but according to their lengths —that is, according to the distance intervening from any phase to the recurrence of the same phase— and this may be shortened or lengthened at will by setting the grating in motion to or from the luminous source. It is true this change will always be very small, for it is in proportion to the ratio of the velocity of approach or recession to the velocity of propagation of light. But with the relative velocities that are found among the heavenly bodies it is often far from being insensible. A velocity of approach effectively shortens the wave-length, and the line is displaced by the grating from its normal position towards the violet. A recession displaces it towards the red. Now if we set side by side the analysed light from opposite limbs of the Sun, at his equator, the east limb is approaching us at a rate of about $1\frac{1}{4}$ miles or 2 kilometres per second, and the west limb receding equally fast. This may be made visible in the spectroscope by several different devices. Cornu by the aid of a prism threw the opposite limbs in succession upon the slit, and observed that as he rocked the prism to and fro to do so, the solar lines rocked to and fro about their mean position, while lines of telluric origin remained unmoved. Dunér, choosing a portion of the spectrum where a number of solar and telluric lines suitable for measurement

IV] THE SPECTROSCOPE 59

are mingled obtained by their means refined measures of the rate of rotation in different solar latitudes. The method is improved by the use of photography, and recently at Edinburgh, Cambridge, and still more perfectly at Mount Wilson, the interesting features of the unequal rotation of the parts of the photosphere have been determined. The displacements in question are shown in Fig. 5. The spectrum is shown triple. The middle spectrum is taken from the west limb of the Sun and the top and bottom spectra from the east limb. It can be noticed that while the pairs belonging to the B band are undisplaced, there are found among them lines in which the middle member is displaced relative to the outer members in the direction in which the B band opens out, that is in the direction of the red. The amount though small is easily visible in the original photograph but the process of reproduction obscures it. These displaced lines are solar, and the displacement is a measure of the solar rotation. A fuller account of the results relating to the Sun's rotation will be given in the next chapter.

Another and most important use of the displacement due to relative motion is found in the detection of movements of constituents of the Sun's atmosphere directly towards us, as in the jets and eruptions which take place, usually over spots. These are sometimes, indeed frequently, so violent that no comparison-spectrum is wanted in order to reveal them. Young has recorded cases where the velocities of advance *and* recession exceeded 200 miles per second. Velocities of 100 miles per second are of frequent occurrence. In such cases the line, let us say the F line, which belongs to hydrogen, in place of showing as usual a dark absorption line, would be brilliantly "reversed," that is to say, bright, and would be broken and torn away from its straight path to an amount that might be more than one hundred times as great as the displacement shown by the Sun's rotation. As a rule the length of the slit would exceed in length the highly disturbed area and the torn piece would occupy a part only of the length of the line.

Relative motion producing the Doppler displacement is one only of the causes that produce shifts in the standard position of spectral lines. There are many others. It seems indeed that almost any differences, for example a difference of direction in the electric current which we use for exciting the element to luminous emission, may produce shifts which are not even very small, and which vary from one line to another. We can hardly expect to follow out disturbances of that character. We must confine attention to those which follow simpler lines, for example pressure.

To obtain sharp lines in a spectrum generated

terrestrially the pressure must be kept low. If the spectrum is produced within a tube or chamber in which the pressure may be varied, an increase of pressure in the first instance generally broadens the lines, and increases the intensity of emission. As the pressure increases the broadening increases, promising ultimately to replace the line spectrum by the continuous spectrum shown by incandescent liquids. The broadening is not symmetrical about the original position of the line; it is greater towards the red side and has the effect of displacing the middle of the line in the same way as if the source of light were receding from the observer. It may therefore to some degree confuse a true Doppler effect. It is a very interesting feature of the pressure displacement that all lines of the same element are not similarly affected. For example, in the spectrum of iron they may be divided into groups which undergo displacements about once, twice, or thrice as great as one another, giving an instance of the unforeseeable complexities to which a close study of spectra leads. Yet this close study is essential, for without it we cannot distinguish the lines of one element from another, not to say trace out the effects of relative movement and pressure. Besides these two causes of disturbance we must reckon with a third which is constantly present, namely the cause which gives rise to the feature known as anomalous dispersion. When an absorbing

Fig. 6.

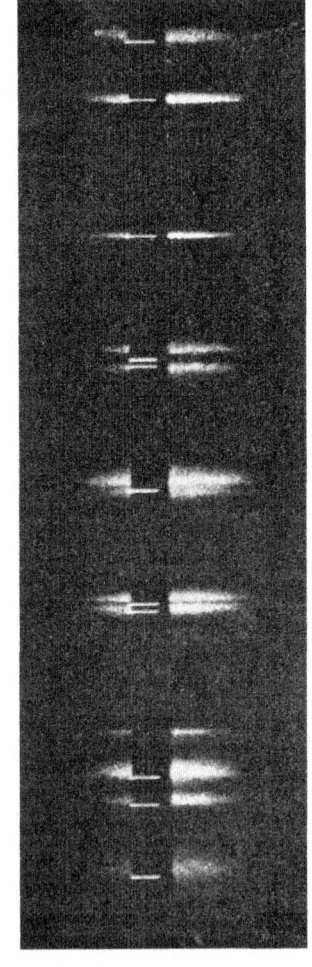

Pressure displacement in Iron Arc (Duffield).

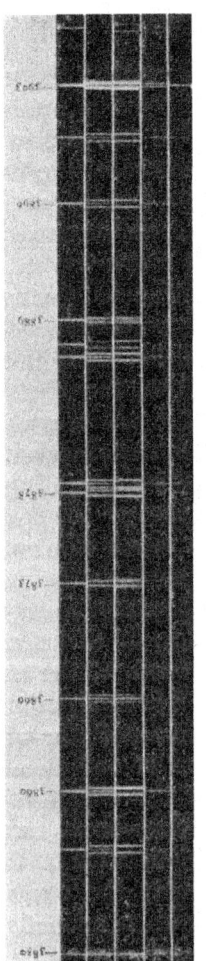

Zeeman Effect on the Iron Spark (King).

system, capable of vibration, is interposed in the path of a wave, the whole must be considered as a single system, and while the absorber has its vibrations increased in intensity, those of the emitting system are also changed, not in intensity only. The period of the vibration which passes on is modified. On the side on which the wave-length is less than that proper to the absorbing system, the transmitted wave-length is diminished ; on the other side the wave-length is increased. The wave-length of the transmitted light determines the refrangibility in respect to a prism or grating. Hence it follows that in the neighbourhood of a place of strong selective absorption the refrangibility of neighbouring lines may be disturbed, the lines upon the blue side suffering a displacement towards the red and those of the red side towards the blue. Absorbing substances are known in which this phenomenon is so prominent that the spectrum is, as it were, cut in two and the colours appear in dislocated order. The best known of them is a solution of fuchsin. But it must not be regarded as an exceptional phenomenon. More or less it is bound to appear whenever absorption takes place, and its effects in common cases have been shown and measured in the laboratory. It must then affect the solar spectrum. Just how and where is a matter that will receive in the future a good deal of attention.

The effect of a magnetic field upon the lines of

a spectrum, predicted by Lorentz and realized by Zeeman, has recently found a brilliant application in the case of the Sun. If the spectrum is generated in a strong magnetic field most of the single lines are turned into doublets or triplets, doublets in the case when they are viewed in the direction of the lines of magnetic force, and triplets when viewed at right angles to this direction. This resolution is in fact a recognition of the mixed character of ordinary light in respect to polarization. The two components of the doublet are circularly polarized in opposite senses, the three components of the triplet are plane polarized, the middle one having a plane of polarization perpendicular to that of its two neighbours. There are besides many complicated cases upon which we do not touch at the moment. The opposite polarizations can be recognized by the help of a Nicol prism, and as will be described in somewhat more detail in the next chapter, this test has been used by Hale to render visible magnetic fields in connection with Sun spots.

CHAPTER V

DESCRIPTION OF THE SUN'S SURFACE

THE Sun's surface shows a rotation, in the same sense as that of the Earth, so that points which appear upon the east limb move across the disc and disappear upon the west. The motion was first found by watching the motions of the spots, and though observations of spots are full of anomalies the position of the Sun's equator and axis of rotation is still found in this way. His equator is inclined but little—7°—to the ecliptic and its plane cuts the plane of the ecliptic in longitude 75° and 255°, that is to say when we look at the Sun on June 6th or December 6th we see his equator projected as a diameter across his disc, crossing west to east from below the ecliptic to above, on the former date, and from above to below on the latter. Half way between, namely on September 8th and March 8th, we look upon the Sun's equator as an oval running 7° from his centre, below the centre upon the latter date and above the centre on the former. Spots circulating about the disc will follow parallel tracks, according to their latitude. As to the rate of rotation, this too was first found reliably from observations of the spots and from these emerged the remarkable fact that the whole does not rotate together at the same rate. The rate falls off as we pass from the equator towards either pole, so that in latitude 60° the time taken for a rotation is greater than that at the equator by nearly one-fifth part. That this is a phenomenon in no way belonging to the spots, but to the general solar surface, is brought out by the spectroscope which reveals the minute Doppler displacement of lines taken from the limbs, due to the advance or recession in consequence of the rotation of the Sun's globe. If this displacement is followed in different latitudes, it tells the same tale even more reliably, for though the quantity to be observed is very small, it is free from the large irregularities that confuse spot observations. The observations, as originally made by Dunér, were visual and relied upon a micrometric comparison of four lines, about $\lambda\,6304$, of which two were solar and two telluric, the latter of course being undisplaced. To-day the same observation is better made by photography, and by means of prisms that reflect simultaneously upon the slit the two limbs of the Sun, the telluric lines can be dispensed with and one displacement made to serve as measuring mark for its opposite. By this means the investigation can be carried on in any part of the spectrum and the behaviour of lines belonging to different elements may be studied. This is important because different

v] DESCRIPTION OF THE SUN'S SURFACE 67

elements are found at different strata of the Sun's atmosphere. Thus hydrogen is found at a very high level and lanthanum at a low one. So far only a beginning has been made of this study, at Mount Wilson Observatory, but it may be looked to in the future to tell us whether the rate of rotation increases or decreases as we rise from lower to higher levels of the Sun's strata. Until this is clearly known any explanation of the meaning of the unequal surface motion must remain speculative. But it may be remarked that it would present little difficulty if the rotation is greater within; for then imagining for simplicity only two strata, the inner one could be flattened relatively to the outer one and would approach it more nearly at the equator than elsewhere, and so would impart by friction progressively more of its greater motion to the outer stratum as the equator was approached. Belopolsky made an experiment in 1886 upon these lines, rotating a globe of glass filled with liquid. When the whole was moving as one piece the globe was stopped, thus exerting a drag upon the liquid; and thereupon particles of stearin which were in suspension in the liquid revealed much the same motion as is noticed upon the Sun's disc. They drifted away from the equator with continually decreasing velocity of rotation, ultimately completing their circulation by turning inwards towards the axis. From this point of view

68   THE SUN   [CH.

the Sun's peculiar law of rotation is merely an expression of the fact of a fluid interior, with angular velocity increasing inwards.

The Sun presents a disc to view which is noticeably darker at the edge than at the centre. It is even more noticeably so in a photograph, for its

Fig. 7. Langley's Drawing of Sun Spots and details of Solar Surface.

deficiency is greater in photographic than in visual light. Besides this darkening, which is due to general absorption in the Sun's atmosphere, the surface under suitable magnification appears mottled all over. Generally speaking these mottlings are

v] DESCRIPTION OF THE SUN'S SURFACE 69

seen with better detail by eye than in a photograph, for the eye can pick out the moments of perfect seeing when the atmosphere for a few seconds is perfectly steady. By collecting the impressions of such moments as this S. P. Langley produced the drawing of which a reproduction is shown in Fig. 7. Rice grains upon a greyish cloth is the usual description of them. The reproduction exaggerates the contrast. The ground is of course not really grey, but it is less intensely brilliant than the clouds or "grains." These clouds are in rapid, even furious movement, across the surface. To prove this we must trust to photographs taken in rapid succession. As a rule, however rapid the succession, no identity of shape or position can be recognized, but Hansky at Pulkowa succeeded in obtaining four photographs of the formations in the neighbourhood of a small spot, at intervals of a few seconds, where the shapes of the clouds could be told again so that their movement could in some degree be determined, and it pointed to a state of extreme agitation.

Up to the present we have referred many times to the Sun spots but have not given any description of them. In their superficial appearance they are represented very beautifully by Langley's drawing, which is more faithful and exact than a photograph and depicts what may be seen momentarily at brief intervals of perfect seeing. With a moderate telescope there is nothing more interesting to watch because of the detail they present and because of their strange groups and varying forms. They arise apparently from nothing and in the course of not many days may attain to diameters of many thousand miles. They always have a dark centre or umbra, by contrast intensely dark, but only by contrast. The impression is that of looking into a black cavern, but there is nothing to confirm the idea that the umbra is really a hollow. Between the umbra and the general face of the Sun's disc is the penumbra, a more or less complete ring, darker than the disc but much brighter than the umbra. Formerly observations used to be adduced of perspective effects in the penumbra as the spot approached the limb, by which the penumbra upon the farther side came relatively more into view than that upon the nearer, as it would do if it were a shallow depressed lip surrounding a hole. But these are now discredited. The shape undergoes such great changes in the course of the days that are required to carry them across to the limb that it is next to impossible to say, even upon the average, that there is any such real effect. For example, a spot will be carried into view upon the east limb showing apparently a more visible penumbra upon its eastern edge, as the theory of depression would require; but when it comes fairly upon the disc it will perhaps be

DESCRIPTION OF THE SUN'S SURFACE

clear that the effect was not a perspective effect at all, but represented a real difference of no significance between the breadths of the penumbra upon the east and west. No spots are long lived. They may spring up and die out in the course of the fortnight that includes a half-revolution of the globe. They may survive a whole revolution or even two revolutions, reappearing again after being carried away on to the invisible side of the body. They become effaced by the photosphere bridging them over, their symmetry of shape is lost and the great spot is broken into two parts, which affect one another somewhat, and these again are broken up and drawn out into ragged strings of small spots. They are confined almost exclusively to two zones of north and south latitudes, namely between about 60° and 20°. They are unknown at the poles and rare at the equator. As everyone knows they have successive periods of plenty and of scarcity. This will be discussed in detail in the next chapter. It must be confessed that up to a very recent date writings upon the Sun spots and upon the Sun's surface generally were of an unsatisfactory character. Differentiated detailed knowledge which would apply to other than special cases was not to be had, and in its absence the subject became a prey to speculation and to those who loved to "explain" one obscure phenomenon in terms of another more obscure still, and who considered that because scientific men were for the time being at a standstill, the opportunity had arrived for popular oracles to step in. It may be said that the spectroheliograph has spelt the end of all this by restoring a normal rate of scientific progress.

The difficulty arose from two causes—the Sun spots and any other features belonged to special strata of the Sun's body, but these and other strata certainly intermingled with one another and reacted upon one another, and it was not to be expected that the behaviour of the visible strata could be understood without the invisible; and again, the spots themselves were certainly intricate phenomena, as difficult to describe as the continent of America, offering too many features to the attention, and no single dominating one as a key to the rest. The faculae and prominences were in just the same position. All was detached, too unconditioned for theory, too incomplete for knowledge.

The spectroscope provided the means of isolation required, and it promises to supply us in due course with information about the whole solar atmosphere and the distribution and movements of the materials of which it is composed. In a spectrogram each line is a record of the presence and the state of a separate chemical element at the spot of the disc to which the slit is directed. If this record could be read v] DESCRIPTION OF THE SUN'S SURFACE 73

for that special line for the whole disc, we should have the same information summed up for the whole Sun. This may be done with an apparatus which, in essence, is comparatively simple. Let the light

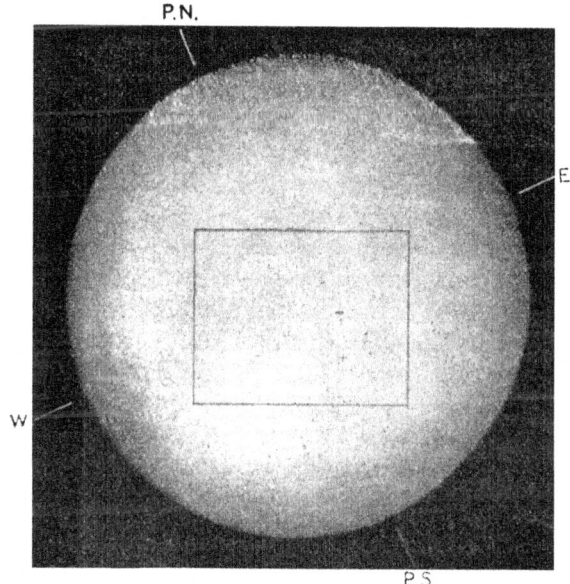

Fig. 8. Direct Photograph of Sun (Meudon), March 21, 1910.

only from the line in question be allowed to pass to the photographic plate, by means of a second slit, at the focus of the camera, the jaws of which shut off

74 THE SUN [CH.

all the rest of the spectrum. Let both the first and second slits be long enough to extend right across the image of the Sun. Move the image of the Sun across the first slit, then the light which passes

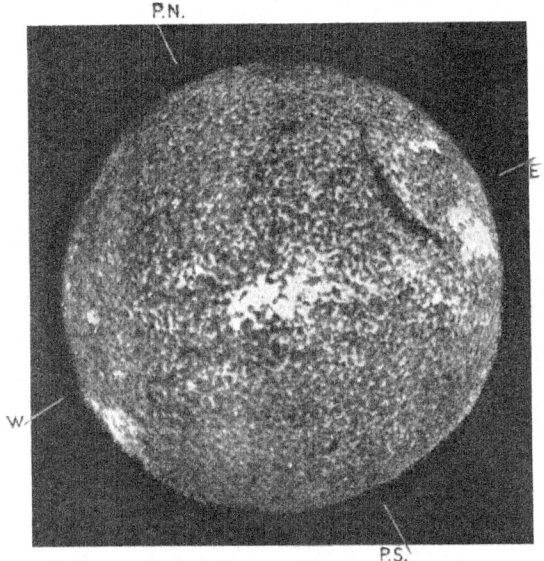

Fig. 9. Photograph of Sun in Calcium Light (K_3), (Meudon), March 21, 1910.

through the second slit will come at every moment from different strips of the Sun's surface, and if the photographic plate be moved behind the second slit,

v] DESCRIPTION OF THE SUN'S SURFACE 75

in unison with the movement of the Sun's image across the first slit, a record will be given, not of the radiations of every substance mixed together, as in

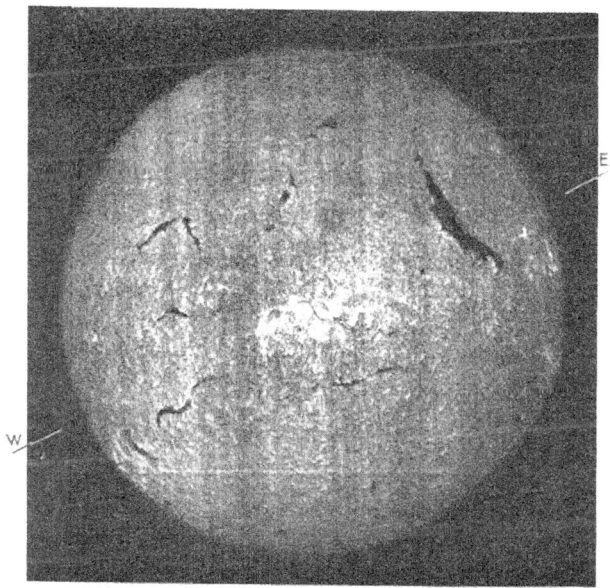

Fig. 10. Photograph of Sun in Chromospheric Layer of Hydrogen Light (H_a), (Meudon), March 21, 1910.

ordinary photographs or visual observations of the Sun's disc, but of the states of some isolated substances as hydrogen or calcium, and even of different strata of these. A difficulty will occur at once. Sodium, let us say, is represented by two strong lines, but they are dark, and if we set the second slit upon one of them it would give us presumably a dark sun, and nothing else, the emission of energy to which we are trusting to affect the photographic plate being trapped upon its way outwards by the sodium vapour of the Sun's atmosphere. But this is not universally the case. It has long been known that hydrogen lines, especially $C(H_\alpha)$ and $F(H_\beta)$, were seen at special points of the Sun's disc, over spots and elsewhere, brilliantly reversed. In 1891 Hale and Deslandres, independently and almost at the same time, discovered that the broad calcium lines, H and K, at the edge of the ultra-violet, showed the same phenomenon, showing even upon occasion a double reversal, the broad hazy dark line being cut into two halves by a narrow bright one, and this again showing a still narrower dark line along its centre. Doubtless, as Deslandres maintains, these represent different layers of the Sun's atmosphere, the original hazy dark line representing calcium at the lower stratum which is answerable for most of the absorption lines of the spectrum, the bright line representing the calcium of the chromosphere as we see it in some prominences, and dark line across this a still higher level.

Following this discovery Hale and Deslandres,

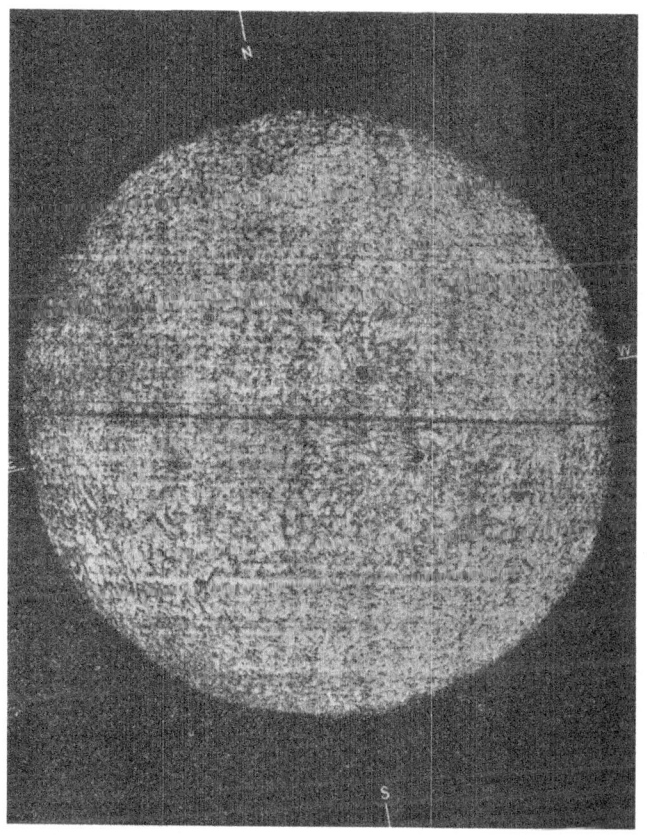

Fig. 11. Photograph of Sun in Hydrogen Light (H_α), showing vortices surrounding Sun Spots (Mount Wilson).

78   THE SUN   [CH.

again almost simultaneously, realized the spectroheliograph, or instrument, of which the essentials

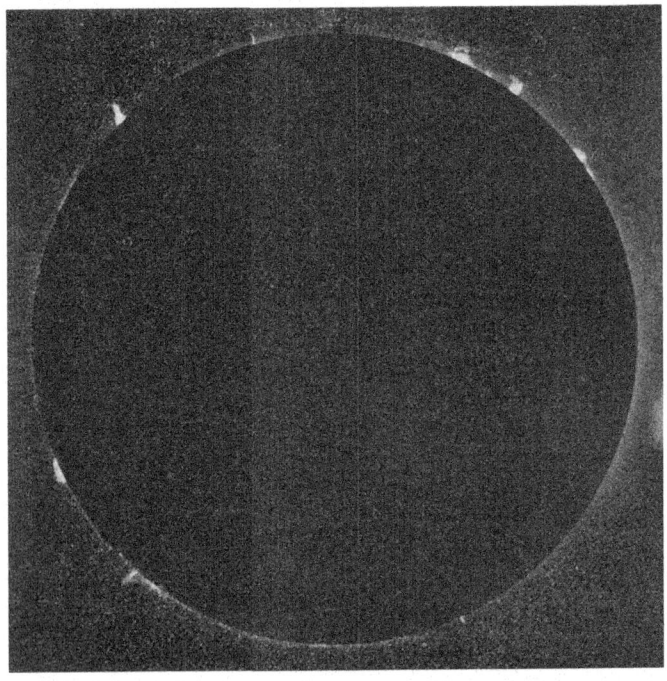

Fig. 12. Photograph of Prominences (H_α) with Sun's disc screened (Mount Wilson), September 20, 1909.

are described above. The second slit is set, let us

v] DESCRIPTION OF THE SUN'S SURFACE 79

say, upon the middle of the K line, and a photograph taken as described above. We then get a picture of the Sun which is bright where the line is reversed and dark where it is unreversed, and is in fact a map of the distribution of the higher clouds of calcium over the whole of the Sun's disc. Such clouds recall the faculae or brilliant regions which were well known in the neighbourhood of spots. They are called by Hale flocculi. The faculae are indeed very often large flocculi and nothing more. Still more detail may be obtained, with a still better separation of the strata, by using, as Deslandres has done, the second slit so narrow as to allow only the passage through it of the double reversal at the centre, or the first reversal on the red or the violet sides. In ordinary cases the breadth of this slit would be of the order of one-thousandth of an inch and the length of exposure would be proportionately great.

The illustrations show in a sufficiently striking way the fruits of this discovery and the great additions it has already made to our detailed knowledge of the Sun. The most brilliant of these is Hale's discovery of hydrogen vortices surrounding many Sun spots, with the accompaniment of strong magnetic fields. These vortices or cyclones follow generally the same rule as obtains in cyclonic circulation upon the Earth, those of the northern hemisphere making their revolution counter-clockwise and those of the southern in the reverse direction. But to this there are exceptions.

Intimately connected with these cyclonic motions was Hale's discovery of strong magnetic fields in Sun spots. The Zeeman effect was the means by which this was detected. It has been described in the last chapter. When the field is less intense than that which produces a clear separation of the line into components, it is widened, and the two sides of the widened line, representing two imperfectly separated components, are oppositely polarized. Hale examined the widened lines of Sun spots with a Nicol prism, and found that the right or left-hand edge could be cut off at will, by rotating the prism. It is supposed this magnetic field is produced by vorticose motion of negatively charged ions, moving inwards towards the centre of the spot. But this is travelling beyond observation. With respect to the actual circulation in spots, Dr St John, confirming and extending observations of Mr Evershed, finds the lower regions of spots, on the level of the reversing layer, are the seat of a radial outflow of many different substances, among which iron is prominent, while there is a corresponding inflow at the higher level of the chromosphere, of which calcium and hydrogen are the chief constituents. But the inflow does not compensate the outflow in either mass or energy.

Outside the spots slight but apparently clear traces of a magnetic field have been detected by Hale and St John over the rest of the disc.

Not less interesting than the vortices are the features discovered by Deslandres and known by the names of "filaments," or in less pronounced cases "alignments." These come out in perfection only with the use of a very narrow slit, separating the different parts of the H_α or K lines from one another and thus recording distinctly what exists at different levels. These filaments and alignments form a species of network over the disc, enjoying the same degree of permanence as the spots, that is to say, occasionally reappearing after a rotation of the disc. They belong to a higher level than the spots. As the slit is set in succession from the border towards the middle of the line, the spots which are shown visually or by direct photographs of the surface, are encroached upon and finally disappear altogether, and in the same succession the *réseau* of filaments comes out.

To understand the circulation of the Sun's atmosphere is at least as important as to recognize formations existing in it. Indeed these latter partly owe their importance to their use as measuring points for determining the former. We may study it by comparing successive photographs taken after the lapse of a few days or hours, or of a few seconds as Hansky did, but much the best way is an instantaneous record of velocity. Such records can be obtained in the line of sight, and Deslandres has devised a variety of the spectroheliograph which gives this information. The idea is very simple. Let the displacement of the K line be recorded over the whole of the disc. When this is done it will show in what regions calcium vapour is ascending and in what regions it is descending, whether in wide reaching surges or in isolated jets. The amount of information contained in such a record is enormous, and we can look to it confidently in the future to increase or rather to create the precise knowledge of the circulation of the Sun's atmosphere. It has already revealed among other things that the filaments are seats of special disturbance.

CHAPTER VI

PERIODICITY OF THE SUN

IF statistics are collected of the occurrence of Sun spots they reveal a periodical fluctuation which is connected in a most intimate and yet a most baffling manner with other solar, and also with purely telluric phenomena. To reveal the fluctuation is easy. An accurate list kept in almost any manner will show

VI] PERIODICITY OF THE SUN 83

it—whether we take the zones of latitude in which the spots occur, or the total area which is covered by the spots that are visible at any time, or the number which any one observer with any special instrument may count, or the number of groups of spots visible, or finally any artificial numbers put together by combining information obtained upon such different bases. But it is a strange fact that Sun spots were observed and enumerated more or less assiduously by a series of observers for over two hundred years, so that over this long period an almost continuous history of their occurrence may be pieced together, without the opinion ever gaining hold, or even ever being expressed by more than one man—Horrebow in 1776—that they followed any law of statistical regularity. The man who effectually revealed it, Heinrich Schwabe of Dessau, an amateur in every sense, having released himself from trade to devote his time to his two passions of botany and astronomy, began in 1826 a series of counts of Sun spots which gradually became demonstrative of the true character of their fluctuation, and by 1843 enabled him to point clearly to a period which he put at about 10 years; and each subsequent year of his observations brought out the conclusion more clearly. A search through past records, in books, journals, and the proceedings of Societies enabled Wolf of Zürich to carry back the history to a date very little after the first discovery of spots in 1610, and with clear continuity from 1750 onwards, and to show that the consistent average period was $11\frac{1}{9}$ years, but that both heights of successive waves and their lengths were subject to large variations. These declare themselves very clearly in the diagram. The correct presentation of the fluctuations shown in this diagram, and their analysis, first arithmetically, but also in their physical relations, is the essential problem of the Sun's periodicity.

The material from which Wolf constructed his diagram was of the most heterogeneous kind possible —casual observations of important spots and groups, systematic surveys, by experienced and by inexperienced observers, with large and small instruments, and so forth—and he reduced them to one footing in a manner which was necessarily arbitrary; that is to say, he formed a representative number for each month out of the formula $r = k(f+10g)$ where g is the number of groups and isolated spots, and f is the number of separate spots that can be counted including those on the groups, while k is a factor depending upon the estimated efficiency of the observer and his instrument. All this is highly arbitrary, and as it is clear that any treatment of the ancient observations must be arbitrary it becomes an important question how far the features of the curve are thus arbitrarily produced. An answer can be derived by making a

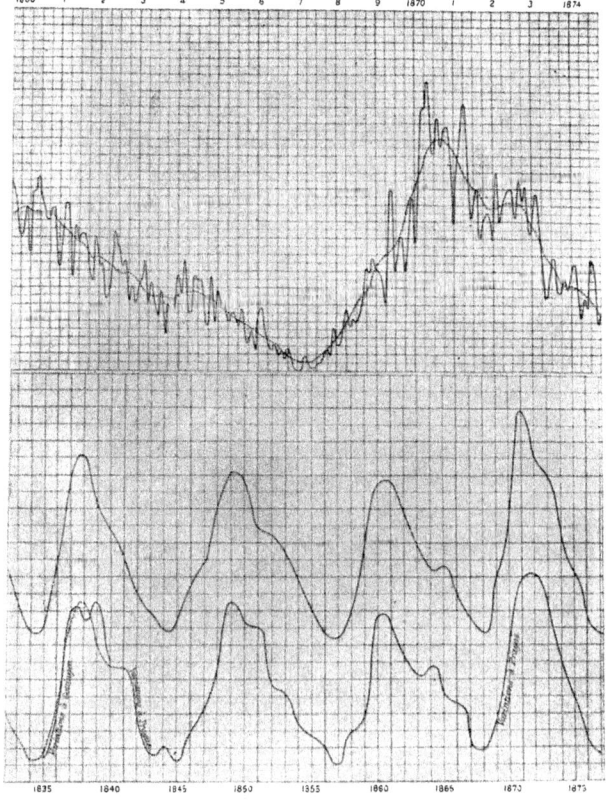

Fig. 13. Portion of Wolf's diagram of Sun Spot Numbers and Magnetic Variation.

different estimate in the modern observations, which shall be as free as possible from this objection. At Greenwich a collection is made of photographs of the Sun upon every available day, and the relative area covered by spots is summed up, allowing properly for the foreshortening. The days when a photograph is impossible are made out by contributions from India and (until recently) from Mauritius. This estimate of the Sun's activity may be considered wholly free from personal bias. Upon a somewhat similar plan it has been carried back by means of Schwabe's and Carrington's drawings of the Sun to 1832. The comparison of these measures is, on the whole, decidedly favourable to Wolf's relative numbers—artificial as the latter may seem in construction. Indeed they appear more homogeneous than the estimates of area, because the latter are unavoidably troubled by three or four changes of plan. A factor can be assigned—upon the average it is 12·5—which will convert pretty closely Wolf's numbers into the millionths of the area of the disc in which the others are expressed. We may consider then that Wolf's representation of solar activity is well justified, and can feel confidence in the undulations which, at the cost of applying immense learning and industry to the old and mostly forgotten records, he has carried back to 1750. But there is another fact that gives substance to Wolf's curves, especially where they

PERIODICITY OF THE SUN

rely upon old observations. Old observers were probably as acute and careful and systematic as modern ones, but their instruments were defective and they had no plan; and this might seem a weak feature in Wolf's record were it not for the wonderful way in which it is borne out in every feature by the curves of magnetic variation which Wolf also constructed. That solar activity was mirrored in magnetic activity upon the Earth was totally unsuspected till 1852, when Sabine, Wolf and Gautier independently showed it to be the case. It was indeed easy to show, when the Sun spot curve emerged; but Wolf's industry again found a field to work in carrying the record back, and the results are such that we might almost read the one curve for the other. Together they leave no room for doubt that Wolf's numbers, right back to 1750, give a true history of solar disturbance, whatever may be the weakness, in places, of the material upon which he was obliged to rely.

When the curve that expresses the solar oscillation is found we have the problem before us to decipher the true character of its variation, as to periods, and varying intensity, to ascertain the real nature and the extent of its intimate connection with terrestrial magnetism, and then more generally its relationship with other phenomena, whether solar, terrestrial, or cosmical.

At a very early stage Wolf fixed the general period at 11⅑ years, and in this there is agreement; Newcomb's revision gives 11·13, and Schuster's 11⅛. But a mere glance will show that such a statement is only a first approximation towards describing the phenomenon. The successive maxima are very unequal in intensity and the intervals between them range between 7·3 years and 17·1 years. If the Sun spots stood alone one would be inclined to write them down as the kind of phenomenon in which no regularity could be expected—transient, sporadic, probably secondary in character, like the outbreaks of an epidemic due to some unknown infection. But the punctual agreement of their curves with those of the magnetic variation forbids us to dismiss so lightly their irregularities. Either magnetic changes are expressly due to Sun spots or both are more or less simultaneous expressions of some common cosmic cause. The latter seems the hypothesis that deserves to be first entertained, and it imposes upon us a strict and general analysis of the curve of change. This is a decidedly difficult problem; for though intricate records like those of the tides or the Moon's longitude which present superficially in some respects no more regularity, have been analysed into their elements with complete success, our command over them has been due to knowing in advance the *periods* at least of all the waves of which the whole is made up, and in that respect we are without knowledge as regards the Sun spots. The main period, even with one or two of its harmonics, is inadequate alone to explain them, and we are left with the task of bringing to the surface subordinate fluctuations, which may be numerous and complicated, by trial.

The only period which inspection suggests with any plausibility is $33\frac{1}{3}$ years, giving three more pronounced maxima in each century; such may be recognized in 1768, 1804—though this was preceded by a much larger one—1837, 1871, 1905, and it is truly remarkable that the same special prominence, thrice a century, may be traced in ancient Chinese records, which contain notices of great spots seen with the naked eye, throughout some fourteen centuries beginning in the second of the Christian era. This has been brought out by Schuster from a study of Mr Hirayama's record. It suggests some connection with the Leonid meteors and Tempel's comet which run in the same period, and the suggestion should not be rejected because there is at present nothing else known to substantiate it. But apart from this 33-year period, it seems almost that, accident apart, the Sun spot curves might have kept their secret for ever had not a more systematic method been brought to bear upon them. This is the method of Schuster's periodogram, which is laborious indeed, but not laborious out of proportion to the ground it covers and the obscurity of the regions which it searches.

If in an oscillating record, as for example in the record of a tide-gauge, a period is suggested as present, its coefficient can be investigated by a straightforward arithmetical process, but the examination is tied somewhat helplessly to the period that was suggested. The method of the periodogram is simply to suggest every period, or at any rate such a numerous possibility as to be representative of the whole. If we are to find out all that is contained in a complicated document like the Sun spot curve, we must approach it absolutely without bias, and not trust merely to such suggestions as can be gathered from the surface by inspection. Because the record is limited, and affected by errors, no period will fail to give a certain measure of positive indication of presence, but a true constituent period will stand out above the rest exactly as a bright line in a spectrum stands out from a slightly illuminated background; and, as such a bright line is not, strictly speaking, sharply defined, but falls at a steep but finite rate into the background, so the approach to a true period in the periodogram is marked by a rapid rise above the small irregular and unmeaning oscillations that mark the rest. When the Sun spot curve is examined by this method the result is little short of astonishing. The diagram shows Schuster's

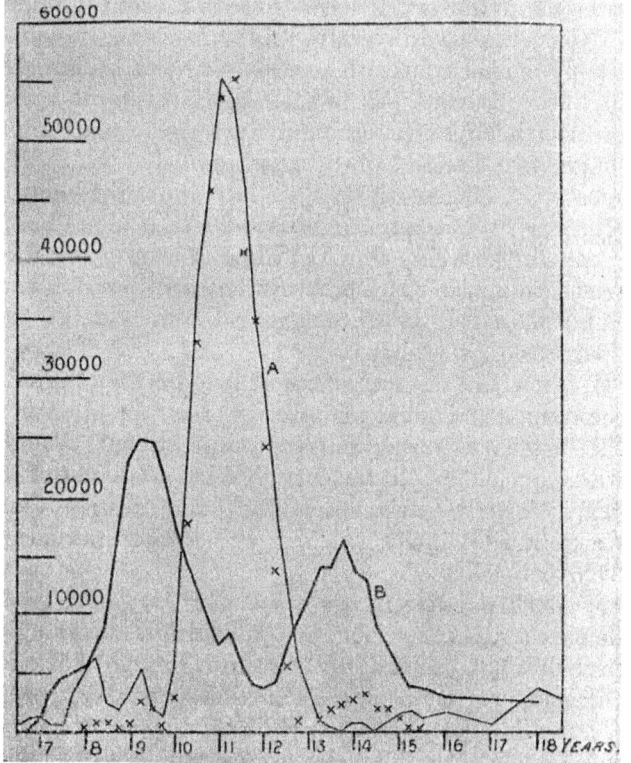

Fig. 14. Schuster's Analysis of Wolf's Sun Spot Numbers for periods between 7 and 18 years.

Curve A: 1825—1900.   Curve B: 1750—1825.

analysis of the last 75 years in curve A and the first 75 years in curve B. If the former were purely and simply a sine curve with period $11\frac{1}{8}$ years it would follow the crosses. But in the latter this period just records itself as present, and the great mass falls under two periods of $9\frac{1}{4}$ years and $13\frac{3}{4}$ years respectively. These appear to have now died down altogether. We might be disposed to attribute the result to the weakness of old observations; but this seems inadmissible, because the repetition of the Sun spot curve by the magnetic curve admits no doubt as to its reality.

When the data are analysed for shorter periods the outcome is not less interesting; for they indicate the presence throughout the whole range, though with varying amplitude, of a period of about 4·78 years. Further, when the whole curve is analysed we obtain an indication of a well-marked period of 8·36 years. Now

$$33\cdot375 = 3 \times 11\cdot125 = 4 \times 8\cdot344 = 7 \times 4\cdot768;$$

hence these three periods are submultiples of a single period of $33\frac{3}{8}$ years, and we are thus brought back to this master-period, as it is sometimes called.

The laborious resolution of Sun spot variations into simple periods has been well compared with the epicycles added one on top of another to explain specific irregularities in planetary motion. When the ellipse was found they were all swept away and

VI] PERIODICITY OF THE SUN 93

summarized under one name and law. Yet they have reappeared, for the ellipse itself required variation, and the simplest expression of its variations consists of a sum of periodic terms, and these are the modern representative of the epicycle which was employed by those who thought geometrically. Similarly there is no simpler way to express what we can find out piecemeal about Sun spot recurrence than in terms of a set of periods.

An objection is sometimes raised to multiple periods in Sun spot activity on the following grounds. The progress from minimum to minimum can be followed, not only statistically by areas but by distribution in latitude. The first outbreaks following a minimum occur in higher latitudes, that is to say, as a rule some 20° south or north from the equator, and thence the activity extends inwards, broad bands of lower latitude being successively affected, until they die out before the following minimum in the neighbourhood of the equator. This is Spörer's "Law of Zones," which indeed was previously remarked by Carrington. The whole follows the general cycle in one large sweep of outbreak and decay, the extreme latitudes being absolutely quiescent after the spots which arose in them have died away. It is contended that if these were not one period but many, such could not be the case. But it cannot be expected that subsidiary phenomena should lie upon the surface, nor that they will be found without search for them. When this is done there is found quite satisfactory evidence of special outbreaks of activity at the maxima of the 4·8-year period, and that in years subsequent to those brought into the calculation from which it was first derived. The same period has been discovered independently in magnetic records. It is hardly a question now of its existence, but of its significance and relationships.

We have so far considered the Sun spot periodicity alone. It is, however, its relationships which enhance so greatly its interest, and first with other features of the Sun.

The prominences are less easy to observe than the spots because their distribution and occurrence cannot be picked up in a moment by means of a photograph; they must be searched out by carrying the spectroscope slit round the Sun's rim. It is, however, certain that their activity generally follows the activity which the spots indicate, the numbers to be found being greater at Sun spot maximum and sometimes falling away altogether during minimum. And a significant fact with respect to the prominences as a measure of the Sun's periodic activity is that they are not confined to the "spot zones"—namely from 5° to 30° south and north of the equator and very rarely beyond—but occur right up to both poles in large though somewhat diminished numbers.

The solar corona again, in a most marked way, changes with the Sun spot cycle. It can only be seen when there is an eclipse and so it is impossible to keep a continuous account of it. But it changes its type with such punctuality that a very experienced draughtsman of coronal forms, on seeing the actual corona for the first time said it was such as he "could have almost drawn with his eyes shut." At Sun spot maximum it is confined, compact, often without outstanding features, and arranged without any marked preference for one direction over another around the whole disc. At minimum on the other hand there is a sharp separation between the neighbourhoods of the poles and those of the equator. From the former for some 30° upon either side of the pole issue fine rays or streamers, not in straight lines but suggesting, whether significantly or not, the lines of force between opposite poles of a magnet, while from the equatorial bands proceed broad and less differentiated sweeps, again not straight and sometimes extending to the immense length of two or three diameters. These two characters appear to recur regularly with the Sun spot cycle.

But though the solar relations of the Sun spot cycle are important as showing how deep seated an indication of solar activity it is, it is its connection to terrestrial magnetism that brings it home to us. It was this connection that first gave prominence and universal recognition to Schwabe's discovery, and we should be at a loss to say which of the two phenomena is the more obscure. There are, broadly speaking, two ways in which the connection may be demonstrated, either by sudden or great outbursts upon the Sun accompanied by magnetic storms upon the Earth, or by the gradual fluctuation of magnetic elements as recorded upon magnetically "quiet" days in sequence with the fluctuations of the Sun spot numbers or any equivalent measure.

The chief use of individual outbreaks is to give precision to our ideas of the relationship. The *locus classicus* is a case observed by Young in 1872. He observed the appearance of a prominence on the west limb of the Sun on August 3. It proved to be over a large spot. It showed disturbances of extraordinary intensity, which he timed. On August 5, while taking the spectrum in the neighbourhood of the same spot, three similar eruptions were noticed, in which the hydrogen lines C and F, where they crossed the spot, flared out sideways across the spectrum "like a blowpipe-jet." On the same dates the automatic records of magnetic declination at Greenwich and at Stonyhurst showed sudden breaks which were so strictly simultaneous with Young's observations as apparently to exclude propagation from the Sun to the Earth short of the velocity of light. If that was the case both must have arisen from some common disturbing cause midway between.

Generally the attempt to bind together individual magnetic storms and individual spot groups yields much less precise conclusions. Mr Maunder has made an examination of 19 great magnetic storms between 1875 and 1903, in connection with the distribution of great spot groups at the time upon the face of the Sun. He concludes that there is undoubtedly a real connection. Several of the storms were of long duration, but all except one are recorded as beginning sharply, and at this sharp beginning there was in every case a great spot group within a region given by one day east of the central meridian or three days west, or upon the average it synchronised with the position of a group about one day past that meridian which passed through the Earth. We may say that this points generally to the tenability of some form of emission theory of influence from the spots to the Earth, but it will not at present permit of more precise statement.

The same question may be approached by another road. Given the records of magnetic storms, they may be examined to see whether their time of occurrence contains indication of a period that must be ascribed to the Sun.

Mr Maunder examining in this way the storms between 1848 and 1903 found that in a number of cases, far too large to ascribe to accident, a storm, after an interval of about 27·3 days was followed by another, and that frequently this recurrence is repeated, three, four, or even six times in succession; that sometimes a rotation or two was overleapt and the sequence was then resumed.

Now this period is that of synodic revolution of the zones of the Sun's surface that contain the Sun spots, so that it reaffirms the conclusion drawn from consideration of the great storms in relation to the great spots, namely that outbreaks of magnetic storms are associated with the approach of special regions of the Sun's surface to the middle of the disc.

The magnetic storms that are the subject of these discussions occur perhaps at the average rate of one per month over a long period—though very unevenly distributed throughout it. We can obtain a far closer and more continuous survey of the relationship if we push the examination to disturbances of a more frequent type. It is the practice at Kew, and elsewhere, to assign to each day's magnetic trace a "character-figure," 0 for a quiet day, 2 for a disturbed one, and 1 for an intermediate one. If there is any real connection between the spottedness of the Sun and the disturbance of the magnetic needle, this should come out by comparing the Sun spot number, or the area, "projected" or "corrected"—it matters little which—with the character-figure. But two cautions

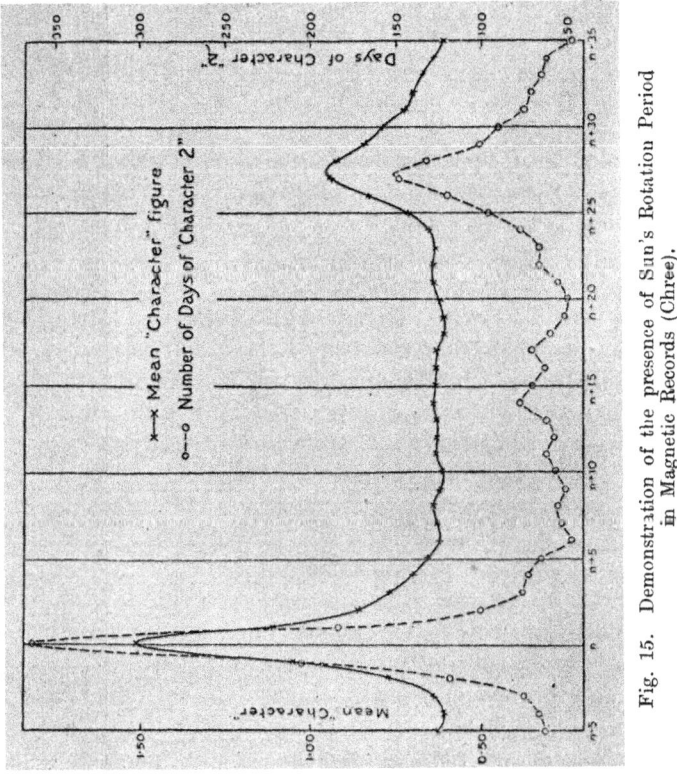

Fig. 15. Demonstration of the presence of Sun's Rotation Period in Magnetic Records (Chree).

must be observed. First both quantities are highly irregular, and it is only upon the average of a great number of cases that any convincing parallelism can be expected, and next it is quite conceivable that there is a connection, but with a time-interval separating the connected numbers. Chree has lately examined the 11 years from 1890 to 1900 from these points of view. In each month five days of character "2," that is disturbed days, are selected, and the character-figure written down for a sequence of preceding and also of following days. The outcome is highly remarkable. When the material is analysed Chree finds that 27·3 days after the occurrence of a disturbed day,—reckoned according to this moderate standard,—there is upon the average a pronounced recurrence of disturbance, while the interval between is a dead level. There is thus a verification, drawn from entirely different material, of Maunder's result.

In a similar way Chree has analysed the records of spottedness of the Sun's disc, its growth and decay upon each side of a large number of chosen days and the corresponding ranges of the horizontal magnetic force. The result is shown in Fig. 16. The average rise and fall of spottedness, about a special day, runs in an extremely smooth and symmetrical curve. The average magnetic range is less smooth but follows unquestionably the same curve, except for an interval or lag of about four days. The character of Chree's

VI]   PERIODICITY OF THE SUN   101

analysis puts these two very definite results in an unassailable statistical position.

It would be altogether premature to draw definite conclusions from these results as to the nature of the connection of Sun spots and terrestrial magnetism, and it would be outside our province to discuss the

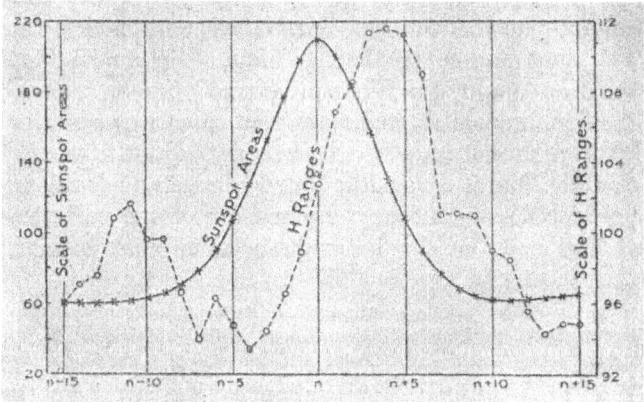

Fig. 16. Comparison of Variation of Spottedness of Sun with simultaneous variations of ranges of horizontal magnetic force upon the Earth (Chree).

latter further. The whole relationship must be left as a thing obviously significant as well as intimate, but mysterious, and only rendered, so far, more difficult, the more we have learnt about it.

Nor is the matter made any clearer by a suspected

still more general relationship. There are stars the light of which varies. Some vary merely because they are periodically occulted by a dark satellite, but with others it is evidently an intrinsic variation; with some it is regular, with others highly irregular, so that their curves present similar problems of interpretation to those of the Sun spots. Indeed one cannot glance over such curves without having the Sun spot curves recalled to mind. Is not in fact the Sun one of the Variable Stars? It is a general opinion among astronomers that the two are akin. Now we do not know why stars vary, so that if we say that the Sun is a variable star we have added this to our problem—that any explanation of the variation of Sun spots or magnetic disturbance must be such that it can be applied also to the stars.

NOTE.—More than one attempt has been made to connect the periodicity of the Sun with the planets. That of Professor H. H. Turner is the latest and boldest. He supposes that the Sun spots are produced by meteors detached by encounters of Saturn with the Leonid swarm and projected to the Sun. Certain numerical relations would fit very well with this idea. But it does not appear that Saturn could ever approach the meteors closely enough to produce such an effect,—indeed not within a distance about half that which separates the Earth from the Sun. Moreover the perturbations of the planets upon the Leonids are already known. Adams calculated them in 1867, and found that Jupiter produces a regression of the node of $20'$ in $33\frac{1}{4}$ years, Saturn $7'$ and Uranus $1'$. These small amounts are known to tally well with observation; and it seems quite certain that nothing so striking as Professor Turner's theory requires can remain to be discovered.

CHAPTER VII

ECLIPSES OF THE SUN

ECLIPSES of the Sun are phenomena which have attracted attention and have been recorded from the earliest period. Indeed they so far predate scientific astronomy itself that they have furnished the name by which the Sun's path is known; for this path is now universally called the Ecliptic, when we pass beyond the merely picturesque conception embodied in the name Zodiac, or as it translates itself into German, *Thierkreis*. Though far from uncommon upon the surface of the Earth generally, their casual occurrence at any particular place is as rare as it is striking. The Sun casts behind the Moon a long pencil of shadow and when the two bodies are nearly enough in line with the Earth this may reach the Earth's surface. Whether it does or no is a question of the relative distances of the Sun and Moon, that is to say whether the Moon shows to the Earth a greater angular diameter than the Sun; for it so happens that the mean diameters of the two are very nearly the same, namely 32′ for the Sun and 31′ for the Moon, with a possible fluctuation of 0′·5 either way for the former and 1′·3 for the latter. It is therefore always a matter of touch and go whether in any case of alignment any spot of total shadow is found upon the Earth; when it is, the greatest diameter it ever reaches is about 120 miles.

Since the Sun is very remote compared with the

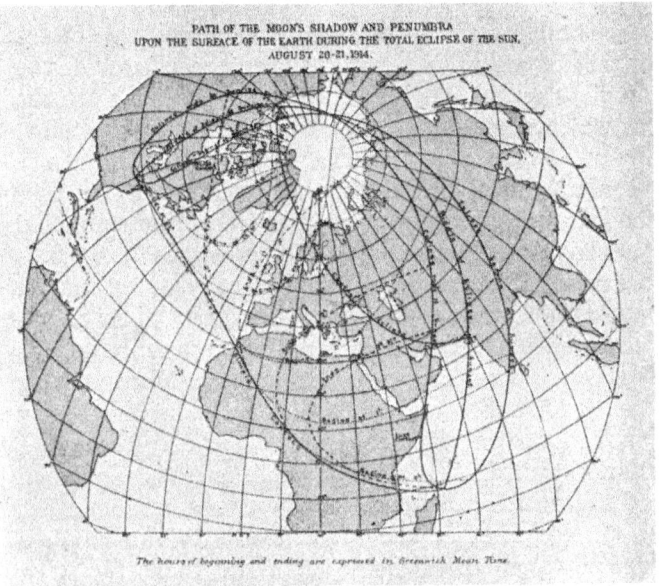

Fig. 17. Chart from the *Nautical Almanac* showing the limits, and line of central eclipse.

Moon, the shadow travels with a velocity which is almost the same as that of the Moon itself, namely at the rate of 60 times the Earth's circumference in

VII] ECLIPSES OF THE SUN 105

$27\frac{1}{3}$ days, or 2·2 times the Earth's circumference per day. Thus while the Earth's surface is travelling from west to east in its daily rotation, the Moon's shadow upon it travels in the same direction, at the least more than twice as fast. In consequence the track of the shadow begins in the west with the Sun just rising, outruns the day and after about six hours of contact with the Earth, ends in the east with the Sun setting. The utmost duration possible at any one place is no more than eight minutes, but very rarely indeed is this maximum approached.

The most admirable of all astronomical cycles is connected with the recurrence of eclipses. This is the so-called Saros which Halley recovered from obscurity or oblivion by rectifying a passage in Pliny's *Natural History*, and identified more by guess than evidence with the Chaldæan cycle or number called Saros. Traditionally the Chaldæans were famous for their skill in predicting eclipses and are supposed to have taught Thales the art in the sixth century B.C., but we know nothing about the origin or method of discovery of this cycle. No doubt it was inferred from lists of recorded eclipses, chiefly of the Moon, for it runs through lunar as well as solar eclipses.

It is easy enough now to reconstruct the Saros in modern terms. The Moon takes 29·531 days to overtake the Sun; the Sun takes 346·6 days to make a circuit back to one of the Moon's nodes, that is, the points when an alignment of Sun, Earth, and Moon is possible. Any close approach to a common multiple of these two periods will give such a cycle as we are seeking, and it may be verified at once that 6585·3 days contains with great exactness 223 lunar months and 19 revolutions with respect to the node. So far it is straightforward, and if it had no further property it would never have asserted itself in solar as well as lunar eclipses, because it passes over the relative distances of the Sun and Moon, which we have seen to be a vital consideration in the character of a solar eclipse. But 6585·3 days also contains 239 revolutions of the Moon in respect to its greatest and least distance, which occupy 27·555 days. Hence after 18 years $11\frac{1}{3}$ days—or $10\frac{1}{3}$ days if five leap years are included—what is substantially the same eclipse track will be repeated. At each recurrence, owing to the odd one-third of a day it will be shifted westward about 120° in longitude, and will only come back after three cycles to approximately the same place upon the map.

So exact is the information given by this great cycle that it is used as the unit of subdivision in Oppolzer's classical *Canon der Finsternisse*, in which he calculates the elements and tracks of every eclipse from B.C. 1207 to A.D. 2161, grouping together all the eclipses of each Saros. The only other cycle deserving

VII] ECLIPSES OF THE SUN 107

comparison with it is the Cycle of Meton, giving the recurrences of the phases of the Moon after 19 years, but that is inferior in respect to the number of quantities which it coordinates, the difficulty of its discovery, its probable antiquity, its significance and its actual utility. The Cycle of Meton was discovered in the fifth century B.C. and it is still alive, for it is the basis of the rule for finding Easter ; but it exists now only in this ecclesiastical survival, while the Saros is a scientific implement we could not easily dispense with.

I understand that it is not at all unlikely that the Assyrian lists of eclipses are still in existence and that archæologists may light upon them at any time. We may safely assume that they were much more extensive than the nineteen lunar eclipses, and one solar, recorded in the Almagest. It is an infinite pity that we have not got them. Supposing they included many of the Sun they would be of untold value for the rectification of the lunar theory, and for other purposes, for they would certainly have been recorded with some care as to time and place. Many eclipses have found mention in mediæval or ancient writings, but except those of professed astronomers like Albategnius or Ptolemy, not a single one of them, standing alone, can be said to have any chronological significance, so little importance did the chroniclers attach to the details and so wanting were they in the systematic curiosity about things of Nature, which is the basis of science. Owing to the sharp limits of time and place under which total solar eclipses are seen, as well as to their striking character, they would seem to offer an ideal occasion for an observer however untrained to supply valuable and precise facts to the astronomer. But they do not. Even an extremely explicit passage of Thucydides is a bone of contention, because he does not state expressly what place his eclipse refers to. Looking at each record separately, as Newcomb said, a kind of fatality appears to have overhung the records. Whether collectively they do not amount to something more is a different question and this will be touched upon again below.

Halley was the first to search for information from ancient eclipses and he found the material too shaky to be demonstrative. If they were exactly given they would be delicate tests of the permanence of the Moon's and the Sun's orbit and motion. Among Halley's learned explorations was the castigation of the text of Albategnius, whose astronomical observations, made at the end of the ninth century, have been available, till very lately, only in a corrupt and ignorant Latin version. A consideration of the eclipses of Albategnius and those found in the Almagest led Halley to the conviction that the Moon's mean motion was not a constant quantity but was subject to acceleration, a thing for which no reason was then to be found. But Halley was unable to prove his discovery because he could not lay down with any precision the longitude of Arracta, the modern Racca, on the Euphrates, where the observations of Albategnius were made. Such are the obstacles in this class of work. After Halley's death the acceleration was derived by Dunthorne from similar material,— but still stumbling at the longitude of Racca,—at the observational value of $10''$ per century—a value pretty generally accepted until recently.

Ptolemy's eclipses are described rather loosely, but at least he was an astronomer. As to whether anything at all can be made of the casual references that have found their way into ancient literature, the best opinions have differed. Newcomb thought not. When an eclipse is mentioned by an author without statement of place, it has been the custom, he says, to write it down as total at the capital of that author's country—the proper conclusion being that its phase is unknown and that it took place at an unknown distance from the author's dwelling, not too far away for him to have heard of it. On the other hand Mr Cowell, accepting the dates and places assigned by chronologists, as best they could, for a number of these eclipses, which seem *primâ facie* to have some pretence to exactness, has shown that a consistent set of small changes in the orbits of the Earth and Moon would bring the best of them remarkably into line. It is true that one of these changes is an acceleration of the mean motion of the Earth about the Sun, for the existence of which no reason has yet been assigned or suggested. But no more was any reason assigned for the acceleration of the Moon which Halley imagined, until, three quarters of a century later, Laplace disinterred at least a part of the cause, in the secular change of the eccentricity of the orbit of the Earth. Knowledge of the Sun's and Moon's orbits is now so exact and so well sifted that any, even a small, change would require a great deal of proving before it was accepted as established; and we have to consider whether material which is separately inconclusive, collectively can possess sufficient force. The whole position at the present moment is not unlike that when Halley wrote, annotating a record of a journey of some merchants to the ruins of Palmyra "...a much greater use of it is, that thereby we are assured that the City of *Aracta*, wherein *Albatâni* made his Observations, was, without doubt, the same which is now called *Racca* on the *Euphrates*. ...The Latitude thereof was observed by that *Albatâni* with great accurateness about eight hundred years since, and therefore I recommend it to all that are curious in such Matters, to endeavour to get some good Observations made at this Place to determine the Height of the *Pole* there, thereby to decide the

VII] ECLIPSES OF THE SUN 111

Controversie, whether there hath really been any change in the *Axis* of the Earth, in so long an Interval; which some great Authors, of late, have been willing to suppose. And if any curious Traveller, or Merchant residing there, would please to observe with due care, the *Phases* of the *Moons Eclipses* at *Bagdat, Aleppo* and *Alexandria*, thereby to determine their Longitudes, they could not do the Science of *Astronomy* a greater Service: For in and near these Places were made all the Observations whereby the Middle Motions of the *Sun* and *Moon* are limited: And I could then pronounce in what Proportion the *Moon's* Motion does Accelerate; which that it does, I think I can demonstrate, and shall (God willing) one day, make it appear to the Publick[1]."

The interest of modern eclipses lies in a region far removed from that of chronology. The glowing disc that we usually see is not, of course, the whole of the Sun. Outside this there is the thin cloak of the photosphere consisting mainly of hydrogen, and bursting up into prominences, and beyond this again the vast extent of the corona. Perhaps we ought to add beyond this the hazy lens of matter that reflects the zodiacal light. No one of these is visible in ordinary daylight because they are smothered in the glare of the Earth's atmosphere. We now know how to outmanœuvre it in the case of the prominences,

[1] *Phil. Trans.* vol. XIX. p. 174 (1695).

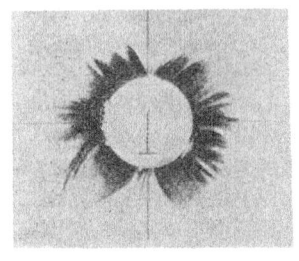

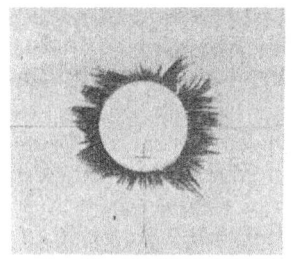
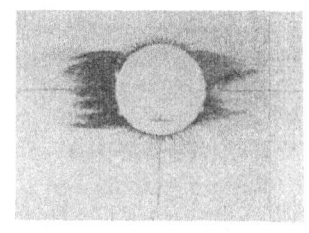

Fig. 18. Hansky's drawings of typical coronal forms, from photographs.

I. 1871, December 12 (Davis); Sun spot maximum.
II. 1878, July 29 (Peers); Sun spot minimum.
III. 1886, August 29 (Schuster); Sun spot maximum.
IV. 1889, January 1 (Barnard); Sun spot minimum.

CH. VII] ECLIPSES OF THE SUN 113

and they can be observed or photographed piecemeal or for the whole disc. As to the corona, Langley, 12,000 feet high, on the summit of Mount Whitney hoped to have reached a point when the air was so tenuous and free from dust that the glare would no longer conceal it, but in this he was disappointed. So far a total solar eclipse is the only occasion that allows us to examine the corona.

The corona is utterly different in structure from the disc of the photosphere. In the first place it is apparently without any definite boundary; its light diminishes rapidly outwards, and we lose it. Again it is not symmetrical with respect to the centre of the disc, though it is roughly symmetrical with respect to the polar axis of the Sun's rotation.

It has a streaked appearance as if made up of filaments; of which the changes during the spot-cycle have already been referred to; at times of Sun spot maximum these are clustered round the disc, very irregularly, with no very marked difference as between the equator and the poles, and the whole visible extent is confined to distance within about one radius' length from the disc. But gradually changing with the spot-cycle, at the time of minimum, there remain for about 30° on either side of each pole the short brush-like filaments that are seen also at maximum, but in the remaining broad band which the equator bisects streamers sometimes of enormous extent, sometimes two diameters, are found. The separation of the regions is quite sharp. These streamers are neither straight nor radial in direction. They suggest somewhat strongly the patterns shown by lines of magnetic force from pole to pole of a magnet sometimes actually running in a loop from point to point of the disc. Of course we cannot say that there is any meaning behind the suggestion, though we know that there exist strong magnetic fields, at least locally, in the Sun spots.

The light from the corona is partly, indeed chiefly, reflected light, as appears from its polarization. But its proper and characteristic spectrum is one of bright lines. In the upper corona these are very few indeed and belong to some element of which we have no other knowledge. But meagre as our means are of investigating it, it may be said that in some respects we know it better than any terrestrial element. For Prof. Nicholson has lately shown that if we imagine an atom built up of negatively charged electrons circulating about a positively charged nucleus, such that the whole is neutral when the number of electrons is five, the oscillations of which it is capable for different numbers of circulating electrons would reproduce the spectrum of the element we call coronium. In the same way he has shown that one which is neutral when the number of electrons is four would reproduce the spectrum of a gaseous

VII] ECLIPSES OF THE SUN 115

nebula. Successful attempts may yet be made to assign similarly the elastic reactions that produce the spectra of known terrestrial substances. Thus hydrogen, which has seemed from Balmer's formula to offer so direct a road for lucky theory into the heart of the inner structure of the elements is now considered as showing chiefly some species of overtones of the same fundamental note, and is turning out, under Prof. Fowler's investigation, to be much more complex than appeared. It cannot be supposed that these analogies of Prof. Nicholson are fallacious; if they are displaced it will not be by proving them erroneous, but by superseding them with an even more far-reaching construction. Thus it is possible that the unknown and perhaps otherwise unknowable coronium may be the right clue for unravelling the otherwise hopelessly tangled web of the structure of the terrestrial chemical elements.

As we go nearer to the Sun's limb more lines are found in the spectrum of the corona.

First we get helium, then hydrogen, and finally the whole normal solar spectrum flashes out in bright lines. For a distance of about five seconds of arc beyond the limb, which gives the continuous spectrum, there extends this layer of incandescent gases, thin enough to give a bright line spectrum. They are in fact the screen which produces the Fraunhofer lines across the otherwise continuous spectrum of the disc, and from that point of view this thin layer might be called the most valuable part of the Sun, for without it we should know next to nothing of the Sun's nature, composition, and internal motions.

The "flash" spectrum, like the corona, is always present but cannot be seen outside of an eclipse because of the diffusion in the atmosphere of light coming from the much intenser disc; but unlike the corona, it does not require a total eclipse to exhibit it. At a partial eclipse in 1912, when the Sun was reduced to a very narrow crescent, the horns of which turned round the disc as the Moon moved across its face, it was noticed by Prof. Fowler and by Prof. Newall that just at the tip of the horns the bright line spectrum could be picked up. Formerly it had only been seen for a few seconds at the beginning and end of each total eclipse. The difference is important because by going just outside the line of totality in a total eclipse the cusps showing the flash spectrum may be followed for an hour or more, so that leisurely arrangements, high dispersions, gratings, and narrow slits—all that goes to make exact the observations of the spectrum of the limb—may be brought into service.

The last phenomena which are always to be seen in a total eclipse of the Sun are the Prominences, but as is well known, these may be seen at any time by searching the outskirts of the limb, till a place is found where one of the hydrogen lines, say H_a, that is Fraunhofer's C, appears bright, and then opening the slit till the whole figure is seen. It is only historically that these are specially connected with total eclipses, since it was these that first revealed the prominences, and suggested to Lockyer and Janssen the procedure necessary for showing them upon any day.

Astronomers have little to reproach themselves with, in regard to the zeal with which they have observed total eclipses. There is in the Observatory of Meudon a picture which represents Paris at night invested by the Prussian lines. A balloon is floating in the sky and in this balloon was Janssen, making his way out of the beleaguered city to observe the solar eclipse of 1870. Unfortunately that eclipse was cloudy. Newcomb tells how the late Father Perry suffered so much at sea on one expedition that he vowed he would never go upon another, but was next met on his way to Kerguelen Island, in the stormiest ocean on the globe. Shortly after, Perry lost his life at the West Indian eclipse of 1886, where he perished of fever. Newcomb himself travelled over land and water, chiefly in canoe, for six weeks, to see the eclipse of 1860 in North West Canada. It again was cloudy. Among more recent examples, we have had Labrador, Madagascar, Sumatra, Easter Island, to mention only the more out-of-the-way spots, all provided with observers.

To the uninitiated the whole seems so costly and precarious that difficulty is felt in understanding just why so much pains are taken, and it is worth while to indicate the astronomer's point of view. It is, and always must be, precarious in point of weather; and that is the reason why every possible point is occupied, even those most out of the way. There is however a pretty narrow limit to distance that can be travelled overland, away from railways, with heavy and delicate instruments, and occasionally, as last year in Brazil, where it was not feasible to choose different and good sites, all the astronomers of all countries were sent empty away by the same clouds. But for the rest weather is the only unmanageable factor. During the past twenty years the Sun has been eclipsed for perhaps half an hour in all, but it is not an exaggeration to say that every second has been turned to account. An eclipse camp represents probably the greatest concentration of thought and activity that is anywhere to be seen. The two minutes or three minutes of duration are plotted off in advance, so many seconds to each exposure with each telescope, so as to secure longer and shorter exposures, with different plates with each instrument. Helpers stand on either side of the astronomer, handing him the fresh plates for exposure and receiving those he has exposed. Seconds are counted in a loud voice from the first signal given—the disappearance of the

VII] ECLIPSES OF THE SUN 119

crescent of the Sun, which as the Moon crosses the last portion of the disc, runs rapidly together and vanishes. Before the day all this is rehearsed several times so that there shall be no hitches or surprises. There are unfortunately futile eclipses, but there is little fault to be imputed to the astronomers. Now the Sun cannot be studied properly while we leave one of its phenomena out. If we knew all about the corona, or if there were any other way of observing it, we might be less insistent for the use of every eclipse; but while the case is otherwise, the many failures in the past from bad weather are only an argument to let no possible future opportunity be missed for increasing our knowledge.

CHAPTER VIII

THE SUN AS A STAR

It often falls to the lot of an astronomer to explain his instruments and methods of work to visitors. And if they realize that his "observations" contain almost nothing that readily strikes the imagination but consist in numbers, and numbers without end; that his chief implement is his library, where kindred observations of others are enshrined in large volumes, apparently utterly unreadable and opening an even more limitless ocean of numbers, that this same pursuit has gone on through many centuries, that the greater part of the time and labour is spent upon collating these numbers and calculating from them, and that imaginative construction is somewhat severely banned as mere baseless speculation, then, as a rule they begin to lose their footing and not infrequently they put the question "What is the good of it all?" A short answer and not a bad one is that it is innocent and serves to pass the time; like Corot's picture, "it has never done anyone any harm"; but if this seems wanting in proper seriousness, or requires to be expanded, one may point out that though the labours of the astronomer do not touch life itself and may be ignored without anyone being a penny the worse, yet they have made the frame in which life is set infinitely more august than any unfettered visionary could have made it. When any seer has set himself the task of describing heaven, how childish is the result! One would almost expect to meet Cinderella's coach driving along its golden streets. Compare with it the "vasty halls" of the astronomer, the great, if unsympathetic, scheme which has been forced upon him by the slow force of his calculations, and by him upon the popular imagination. We see that, in a sentence, the great end of astronomy—apart from the essential though trivial services which it renders in getting people punctually and safely from one part of the earth to another and keeping their calendars more or less clear from confusion—is to answer the question, What is the Universe? and that the astronomer has already made great and certain progress towards the answer. We now know with a definiteness that may be called adequate, and, in some respects, is hardly open to improvement, the scale of the solar system. The origin of the parts, the planets and their satellites, may be obscure or speculative, but they are all evidently one, made of the same matter and subject to the same mechanical and physical laws. The next task is to extend our grasp as far as the stars, and in this too a great deal has been done. And, to return to the proper theme of this book, in this task the Sun appears in a special and indispensable relation. It is the nearest star. It is the only one of the stars which is near enough for detailed examination. We must use our knowledge of it, building upon what seems common and omitting what seems peculiar, to form our ideas of the many millions of other stars which the word Universe denotes. We must use it as the key and mark off the doors which it will open from those which it will not.

The question divides itself into three parts under the heads of physical nature, structure, and extent of the stars, covered generally by the character of their spectral type, their relative motions, and their relative distances, with their possible limitation.

The term "spectral type" requires some explanation, because we have up to the present spoken only of the spectrum of the Sun. If a prism is put in front of the object glass of a telescope which is made to follow a field of stars, each of these will be represented in the field of view not by a point but by a spectrum. Of these spectra, as a rule, a proportion—a comparatively small one—will be of a character generally similar to that of the Sun, crossed by a host, more or less, of the same dark lines. The predominant type will be one with no lines except those of hydrogen, but these very dense and broad. And other types may be present which will be particularized below. Now these types are not rigidly separated from one another. Sometimes the hydrogen lines will be accompanied by those of helium, sometimes by solar lines shown faintly, sometimes a decided solar star will show the hydrogen lines with excessive strength, sometimes a host of lines besides those in the solar spectrum. So an attempt is made to arrange them in some sequence which if one pleases may be regarded as merely schematic, but which may possibly represent successive stages of development, through which all, or at any rate most, of the stars, must pass in the course of their life. More than one such scheme has been drawn up, and these differ in detail, for it is hardly possible to proceed without preconceived theory.

THE SUN AS A STAR

I shall confine myself to the one most widely used, namely that of the Draper Catalogue of Stellar Spectra, made at Harvard College Observatory, in the form which it has finally assumed.

The original idea of this catalogue was to start with the type of most frequent occurrence and apparent simplicity, namely that which showed simply the hydrogen series of lines, broad and dark. This was called Class A. Next was put Class B, which besides the hydrogen lines showed a large number of others, which have since appeared as due to helium. Class F showed the hydrogen lines somewhat less broad and intense, while the calcium lines H, K were the most conspicuous feature and the ordinary solar lines were clearly present though on a subordinate plane. Next came Class G of purely solar type, in which the most conspicuous features are the calcium lines H, K and the band G. After Class G was put Class K, in which the light in the shorter wave-lengths is faint, and throughout the spectrum the light is not uniform; the hydrogen lines are fainter than the solar lines proper. Fraunhofer's G, H, K are very prominent together with the calcium line at wave-length 4227. Next comes Class M, which shows decided bands of varying darkness, sharp on the blue side, shaded towards the red. Then we have Class N, again a banded spectrum, but the bands of absorption now sharp on the red side and shaded towards the blue. Then came Class O which showed bright lines, especially those of hydrogen; finally Class P denoted the gaseous nebulae.

After further examination, these have been rearranged into an order which really appears to belong to a sequence of development. We begin with the gaseous nebulae. These contain a gas, unknown upon earth except through Nicholson's theory, and also hydrogen and helium. An organic part of certain nebulae, as for example the great nebulae in Orion, are stars, and these may be supposed to represent the earliest stellar type. Thus we have the Orion stars, which show faint and diffuse lines of helium and hydrogen—that is, Class B; after this, not before it, comes Class A, in which as the helium lines die out, the hydrogen lines attain their maximum intensity; Sirius, and indeed most of the brightest stars are of this type. These are the Hydrogen or Sirian stars. They show white or bluish in the sky, but the name white stars does not distinguish them sufficiently for it includes the next class also, the Calcium stars of Class F, of which Altair and Procyon are examples. The solar stars of Class G have a yellowish tinge; Capella is the type for these. An incipient redness characterises the Class K and is illustrated in Arcturus and more emphatically in Aldeboran.

Beyond this it cannot be said that a single road

is obvious. We have Class M, including Antares, Betelgeux and Mira Ceti, with other variables of long period, showing absorption bands sharp on the side of the blue, and the visually insignificant Class N which shows bands sharp on the side of the red. To which, if to either of the Classes M and N a solar star would degenerate in lapse of time we cannot say. The absorption bands of Class M have been identified by Fowler with titanium oxide, and the same bands have been found by Hale in the Sun spots. Those of Class N arise from hydrocarbons and cyanogen, and cyanogen also has been found in the Sun. Admitting then that the previous path runs fairly plausibly from the nebular stage down to our Sun and beyond, we are somewhat at a loss to assign the final goal. Indeed it is clear that it is by no means necessary that the course of development should be the same for all stars, even when these are pretty nearly on the same apparent footing, and purists have contended that we have as good right to trace the development from the red stars back to the nebula as in the order sketched above, but the balance of opinion inclines to take the progression to be an actual one and typical of the great majority of lucid stars, and thus to give a faithful sketch of the past and future history of our Sun. Yet so reserved are astronomers in adopting conclusions in advance of proof that it recently came as a shock of surprise to find this order of development stamped with a new mark of reality, as it was in a singular way—the importance of which is easier to grasp than the full meaning—from a study of the motions of the stars. To this matter we come next.

The term "fixed stars" is, of course, a misnomer. All the stars are in motion relative to one another, and the Sun among them; and that with speeds so enormous that they make themselves sensible in spite of the meagre means we have for detecting them. These means are two, and give respectively the transverse motions of the stars in magnitude and direction, measured in so many seconds of arc per century, and the line-of-sight motions, measured in miles per second. Each is involved with the motion of the Sun itself. Neither gives the whole story, nor do they altogether dovetail into one another, but because of their differences, they render one another support.

The ideal method of finding the transverse motions would be to get a map or catalogue of the places of the stars at two sufficiently distant epochs and then compare them. The differences of place for each star would be what is called its proper motion in the elapsed time. But with respect to what datum-lines are these maps to be drawn? The equator and equinox which we are compelled to use for reference are themselves in motion, and their motion is much greater than that of the stars, and can only be determined relatively to the stars assumed for the moment as fixed. It is clear that the only thing that may be considered fixed is a mean reference to the whole body of stars. To obtain such a reference, purified from all sources of error, accidental and systematic, is the task which occupies the constructors of star catalogues. Since it brings simultaneously under criticism the theoretical motions of the Earth and the whole mass of observations of stars, since Bradley's time, with telescopes and clocks of very various quality, and subject to errors of which the observers at the time were unconscious, the construction of a fundamental catalogue of star places at a particular epoch with the accompanying proper motions is a work of immense difficulty requiring much critical power. It has however been done, most recently and completely by Boss, and his resulting proper motions of six thousand stars are the basis for any conclusions that are now drawn. It is worth mention that one of these conclusions is that the deduced motion of the Sun relative to the stars is really not very far from that assigned to it, in amount and direction, by Herschel in 1783 upon the ill-founded motions then at his disposal of a mere handful of stars.

It is fairly obvious how a list of proper motions of the stars leads to an inference as to the motion of the Sun. If it appears that, besides a great deal of random motion in all directions, there is any diameter of the celestial sphere towards one end of which the stars as a whole appear to be converging, while as a whole they diverge from the other end of the same, the natural interpretation is that this systematic motion is merely apparent, and that the reality is a motion of the Sun towards the point from which the stars seem to be opening out. Herschel fixed this point, the solar apex at R.A. 262°, Dec. + 26°. Boss gave for the same R.A. 270°·5, Dec.+34°·6. But it is clear that this is not a necessary interpretation. An alternative one is that systematic motion belongs to the stars themselves, or if our knowledge is precise enough we might be able to resolve it into two or several drifts or streams of stars through one another. In spite of the great difficulties as to the material this is the position to which Boss's labours have brought us. They have revealed first, some clusters consisting of say some hundred or two stars, possessed of common motion relative to the rest of the universe. The greatest of these is the Taurus cluster, a group of stars some 15° in diameter, all moving towards the point R.A. 92°, Dec. + 7°.

But this is comparatively a minor matter. A searching analysis confirms what had been already concluded from previously available material, that there is a bipolarity among the stellar motions, not a single polarity, as Herschel supposed, which might be attributed wholly to the Sun. This may be put in several ways, for we are ignorant how much absolute motion to ascribe to the Sun. If we suppose the Sun at rest we must say that there are two streams of stars crossing one another; Prof. Eddington estimates that the numbers involved in these two streams may be half a million, so that it is a phenomenon of an entirely different scale from the common motion of a cluster; if we ascribe such motion to the Sun as will reduce to its simplest expression the remaining systematic motion of the stars, there is still one favoured direction, in the plane of the Galaxy, towards and from which the stars are streaming. There is now no doubt at all about these conclusions, even if we were unable to bring further evidence to confirm them, but they are fully confirmed by investigations of motions in the line of sight. These are free from many of the difficulties that beset the transverse proper motions. In the first place they can be obtained one by one, without a criticism of the whole that remains, and next they give absolute and not merely angular motions which cannot be completely interpreted without knowing the distance also. Their number is as yet by no means so large as that of the others, but they are already numerous and precise enough to confirm the two-stream hypothesis. They do more, for returning to the question of a star's development, they contribute a valuable stamp of reality to the theoretical sequence mentioned earlier.

This they do in a remarkable way, the interpretation of which is itself obscure. If we take the stars of which the line-of-sight motions have been determined and arrange them according to their spectral classes, their velocities are not random as one might have surmised; they are least with Class B, and increase with unfailing progression right up to Class M. Why a star in an early stage of development should move relatively slowly is a matter for speculation, but it is pretty clear from the result that we have got the Classes B to M arranged in their true order.

These wider relationships of the Sun to the universe of stars are open only to very imperfect interpretation until we know much more than we do at present of the distances of the stars from the Sun. If we knew these well, we should know more than the linear scale of things, we should know the absolute transverse proper motions in place of merely the angular values of them, and we should be able to translate magnitude into intrinsic brightness. But progress here is slow, almost out of reach. Though the question of stellar parallax has been in the front of Astronomy since the time of Aristotle, who used it as an argument to disprove a moving Earth, and has been, since the acceptance of

THE SUN AS A STAR

Copernicus's theory, a debt of honour which the science was called upon to pay, we are only now beginning to have enough isolated determinations to supply some idea of how far off we are from adequate knowledge.

If we take a sphere of radius a million times the Earth's orbit, it would include about 18 stars. These are the stars with parallax two-tenths of a second of arc or more. If we go only so far as double this distance, we are near the limit of what our methods of measurement can separate from their own inevitable errors; yet we have hardly touched the fringe of the question. To get anything like an adequate view we require to go at least a hundred times, probably a thousand times as far. It is certain that no means at present available can do this and we are thrown back upon various hypotheses to bridge the gap for the time being, and to give us if possible a tenable view. Such a hypothesis is that magnitude is, generally, an effect of distance; but this seems not to be the case, at any rate in any governing degree. The simplest resource is that of Herschel's counts. By counting the stars that are found in selected areas distributed over the sky, there is found an axis of least density and a plane at right angles to it of greatest density. This plane is the plane of the Galaxy, which is omitted in making the counts. So far, of course, no hypothesis has been brought in.

But if now we make a simple supposition, as that the density of stellar distribution is constant in that part of space where they occur, these counts supply a measure of extent, and we reach Herschel's well-known conclusion, only slightly modified by later knowledge, that the universe forms a figure approximately lens-shaped, but rather flatter upon one side, with the Sun roughly near its middle. The clouds of the Milky Way are not included in this system but they are found ranged about its edge.

Though magnitude cannot be taken safely as a guide to distance, distance must affect magnitude, and one consequence of a uniform stellar density would be the occurrence of approximately four times more stars of a higher magnitude than of a magnitude one unit lower. This can be tested, and it is not the case. The ratio is always less than four, and falls progressively as we get to higher and higher magnitudes.

There are two interpretations of this fact, and as they are not exclusive, it is very possible that each is a contributor to it. The first is that as we go outwards the number of stars actually falls off, so that the universe we know is a gigantic cluster, more scattered at its outward parts. Or again, we may suppose that there is absorption of light in space, not at all owing to the aether, but to scattered material particles. This would have a kindred effect telling most in the extinction of the remoter stars, thrusting

VIII] THE SUN AS A STAR 133

at every stage a larger proportion of them down to a higher magnitude.

With this brief glance at the system of the stars of which our Sun forms a representative, but by no means a prominent, member, we may fitly conclude.

Although, as we foresaw from the beginning, the further we go, the more unsolved problems we open up, it is a legitimate glory of the human mind, if anything is, to have gained what seems a firm footing in knowledge of states, distances, and times all equally impossible to occupy. It conveys the impression which perhaps is not so fallacious or absurd as it seems of something in the mind even more permanent and universal than these. Though the mind is conscious of itself only for one moment, it has the power, perhaps we should say the irony, to see and to say:—

Soles pereunt et imputantur.

BIBLIOGRAPHY

Abbot, C. G. The Sun. 1912.
Albatenius. Opus Astronomicum edidit Nallino. Publ. del r. osservatorio di Brera. Milan. Vol. XL (1903).
Auwers, A. Die Venus Durchgänge 1874 und 1882. Bericht über die deutschen Beobachtungen. Berlin. 1898.
(On the Sun's Diameter.)
Belopolsky, A. Über die Bewegungen auf der Sonnenoberfläche. Astronomische Nachrichten. Bd. 114, p. 153.
Carrington, R. C. Observations of the Spots on the Sun from 1853 to 1861. London and Edinburgh. 1863.
Chree, C. Some Phenomena of Sun Spots and of Terrestrial Magnetism. Phil. Trans. Vol. 212, p. 75 (1912).
Cornu, A. Sur les raies telluriques. Bulletin Astronomique. T. 1, p. 74 (1884).
Cowell, P. H. On ancient eclipses. Monthly Notices of the Royal Astronomical Society. Vols. 65–9 (1905–9).
Deslandres, H. Annales de l'Obs. d'Astronomie Physique de Paris (Meudon). T. IV (1910).
Doppler, C. Über das farbige Licht der Doppelsterne. Abh. d. k. Böhmischen Gesellschaft. d. Wiss. Prague. 5te Folge, Bd. II (1842).
Duffield, W. G. Effect of Pressure upon Arc Spectra. Phil. Trans. Vol. 208, p. 111 (1908).
Dunér, N. C. Recherches sur la Rotation du Soleil. Upsala. 1891.
Dunthorne, R. The Acceleration of the Moon. Phil. Trans. Vol. 46, p. 162 (1749).

BIBLIOGRAPHY 135

Fowler, A. Spectroscopic Observations during Partial Eclipse of the Sun, April 16, 1912. Monthly Notices of the Royal Astronomical Society. Vol. 72, p. 538 (1912).

Fowler, A. Observations of the principal and other series of lines in the Spectrum of Hydrogen. Monthly Notices of the Royal Astronomical Society. Vol. 73, p. 62 (1913).

Fraunhofer, J. Bestimmung der Brechungs- und Farbenzerstreuungs-Vermögen verschiedener Glasarten. Denkschriften der k. Acad. d. Wiss. zu München. Bd. v, p. 193. 1817.

Gill, D. Annals of the Royal Observatory at the Cape of Good Hope. Vols. VI, VII.
(On the Solar Parallax.)

Greenwich. Observations of the Planet Eros, made at the Royal Observatory. 1908.

Hale, G. E., and Ellerman, F. The Five-Foot Spectroheliograph of the Solar Observatory, Mount Wilson. Astrophysical Journal. Vol. XXIII, p. 54.

Halley, E. Emendationes in vetustas Albatenii Observationes Astronomicas. Phil. Trans. Vol. 17, No. 204, p. 913 (1693).

Halley, E. Emendationes in tria Loca Texti Naturalis Historiae Plinii. Phil. Trans. Part 194, p. 535. 1691.

Hansky, A. Mittheilungen der Nikolai-Haupt Sternwarte zu Pulkowo. Bd. 1, p. 81 (1907).

Herschel, W. On the Solar Motion. Phil. Trans. 1783 (Collected Works, I, p. 108); ib. 1805 (Collected Works, II, p. 317).

Hinks, A. R. Solar Parallax Papers. Monthly Notices of the Royal Astronomical Society. LXVIII, LXIX, LXX.

Hirayama, S. Sun Spots in Chinese Annals. The Observatory. Vol. XII, p. 217 (1889).

Humphreys, H. J., and Mohler, J. F. Effect of Pressure on Wave Length. Astrophysical Journal. Vols. 3, 4, 22.

King, A. S. Influence of Magnetic Fields upon Spark Spectra. Publications of the Carnegie Institute. No. 153. Washington. 1912.

BIBLIOGRAPHY

Kirchhoff, G. Untersuchungen über das Sonnenspectrum und die Spectra der chemischen Elemente. Poggendorfs Annalen. Bd. 109 (1866).

Langley, S. P. Researches on Solar Heat. U.S.A. War Department. Papers of the Signal Service. xv. 1884.

Maunder, E. W. Magnetic Disturbances and Sun-Spots. Monthly Notices of the Royal Astronomical Society. Vols. 64, 65. 1904, 1905.

Mount Wilson Observatory Publications: — in Astrophysical Journal.

Newcomb, S. On the Period of the Solar Spots. Astrophysical Journal. Vol. 13, p. 1.

Newcomb, S. On ancient eclipses. Monthly Notices of the Royal Astronomical Society. Vols. 65, 66 (1905, 1906).

Nicholson, J. W. On the constitution of the Corona and on Nebulium. Monthly Notices of the Royal Astronomical Society. Vol. 72 (1912).

Oppolzer, Th. v. Canon der Finsternisse. Denkschriften der k. Akad. der Wiss. (math.-phys. Classe) zu Wien. Bd. 52 (1887).

Pouillet, C. S. M. Sur la chaleur du Soleil. Comptes Rendus, vi, 1838.

Poynting, J. H. Radiation in the Solar System. Phil. Trans. Vol. 202 (1904), p. 525.

Rowland, H. Physical Papers. Baltimore. 1902.

Rowland, H. Preliminary Table of Wave Lengths of the Sola Spectrum. Astrophysical Journal. Vols. 1–5.

Runge, C., and Paschen, F. Oxygen in the Sun. Astrophysical Journal. Vol. iv, p. 137 (1896).

Schuster, A. On the Periodicity of Sun Spots. Phil. Trans. Vol. 206, p. 6 (1906).

Schuster, A. Chinese Records of Sun-spot Periodicity. The Observatory. Vol. xxix, p. 205 (1906).

Schwabe, H. Über die Sonnenflecken in 1843. Astronomische Nachrichten. Vol. 21, p. 223 (1844).

Smyth, C. Piazzi. Trans. Royal Society of Edinburgh. Vol. XXXII, pp. 238, 446 (1883).
Very, F. W. Absorption of Radiation in the Solar Atmosphere. Astrophysical Journal. Vol. 16, p. 73.
Wilson, A. Observations on the Solar Spots. Phil. Trans. Vol. 64, p. 1. 1774.
Wolf, R. Astronomische Mittheilungen. 1860–. Zürich. (Continued by A. Wolfer.)
Young, C. A. The Sun. 2nd Edition. 1905.
Zeeman, P. Phil. Mag. 5th Series. Vol. 43, p. 226 (1897).

NUMERICAL DATA

Sun's Parallax: $8''\cdot807 \pm 0''\cdot0027$.
Mean Distance: $9\cdot283 \times 10^7$ miles $= 1\cdot4939 \times 10^{13}$ cm.
Limits of "probable error" of distance:
$$\pm 3 \times 10^4 \text{ miles} = \pm 5 \times 10^9 \text{ cm.}$$
Variation of Mean Distance:
Jan. 1 $-1\cdot555 \times 10^6$ miles $= -2\cdot502 \times 10^{11}$ cm.
July 1 $+1\cdot555 \times 10^6$ miles $= +2\cdot502 \times 10^{11}$ cm.
Sun's semidiameter: $959''\cdot6$.
$$4\cdot318 \times 10^5 \text{ miles} = 6\cdot949 \times 10^{10} \text{ cm.}$$
Area of Surface: $2\cdot343 \times 10^{12}$ sq. miles $= 6\cdot069 \times 10^{22}$ sq. cm.
Volume: $3\cdot372 \times 10^{17}$ cu. miles $= 1\cdot637 \times 10^{33}$ cu. cm.
Mass: $3\cdot32 \times 10^5 \times$ mass of Earth $= 2\cdot32 \times 10^{33}$ grams.
Density: $\cdot2533 \times$ density of Earth $= 1\cdot400$.
Gravity at Surface:
$27\cdot6 \times$ value at Earth $= 886$ ft./sec.$^2 = 2\cdot70 \times 10^4$ cm./sec.^{2}

Elements of the Sun's Apparent Orbit:
At epoch $1900\cdot0$: $T = 36525$ days.
Obliquity of Ecliptic: $23°\ 27'\ 8'' - 46'''\cdot8\ T$.
Tropical Year: $365\cdot2422$ days.
Sidereal Year: $365\cdot2596$ days.

NUMERICAL DATA

Sun's mean sidereal motion: $129\,597\,743'' \, T$.

Sun's mean longitude freed from aberration:
$$279° \, 41' \, 48'' + 129\,602\,768'' \, T.$$

Eccentricity of Orbit: $\cdot 01675 - \cdot 00004 \, T$.

Longitude of Perigee: $281° \, 13' \, 15'' + 6189'' \, T$.

Rotation:
- Inclination of Sun's Equator to Ecliptic: $7° \, 11'$.
- Longitude of Ascending Node of Equator: $74° \, 35'$ (1900·0).
- Daily angular rotation in latitude ϕ: $10°\!\cdot\!62 + 3°\!\cdot\!99 \cos^2 \phi$.
- Surface Velocity at Equator: 2·06 km./sec.
- Rotation Period: latitude $0°$, $24^{d}\!\cdot\!6$;
  - $\phantom{\text{latitude}}$ $30°$, $26^{d}\!\cdot\!4$;
  - $\phantom{\text{latitude}}$ $60°$, $31^{d}\!\cdot\!0$;
  - $\phantom{\text{latitude}}$ $75°$, $33^{d}\!\cdot\!1$.

Solar Constant, or units of energy received per minute per square centimetre at the Earth's surface: 1·95 calories.

Sun's Effective Temperature considered as a "black body" or ideal radiator: 6000° absolute.

Sun's stellar magnitude $-26\cdot 5$ or 10^{11} as brilliant as a 1st magnitude star.

Apex of Sun's Way: R.A. $270°\!\cdot\!5$, Dec. $+34°\!\cdot\!6$.

Apices of two star drifts:
- Drift I: R.A. $91°$, Dec. $-15°$.
- $\phantom{\text{Drift }}$ II: $\phantom{\text{R.A. }}$ $288°$ $\phantom{\text{Dec. }}$ $+64°$.

INDEX

Abbot, 17
Aberration, 43
Adams, 102
Adiabatic equilibrium, 24
Albategnius, 107, 108
Almagest, 107
Ancient eclipses, 108
Aristotle, 130
Astronomy, aim of, 120
Atmosphere, circulation in terrestrial, 9
—— earth's as heat absorber, 12

Belopolsky, 67
Black Body, 5, 20
Bolometer, 15
Boss, 127
Bradley, 127

Cambridge, 58
Carrington, 86, 93
Chaldaeans, 105
Chinese observations, 89
Chree, 100
Coronae, 113
Corona, 95
Cowell, 109

Deslandres, 76, 82
Displacements of spectral lines
    Doppler, 57
    pressure, 63
    Zeeman, 63, 80

Doppler displacement, 57
Draper catalogue, 123
Dunér, 58, 66
Dunthorne, 109
Dynamics, Law of Least action, 5
—— originating from solar theory, 5
—— Problem of Ibra Bodies, 5

Earth, age of, 27
—— radius as seen from Sun, 33
"Earth's way," 43
Eclipse camp, 118
Eddington, 129
Edinburgh, Royal Observatory, 47, 55
Energy of sun ultimately lost, 18
Eros, 38
Euclid, 3
Evershed, 80

"Fixed" stars, 126
"Flash," 116
Flocculi, 79
Fowler, 115, 116
Fraunhofer, 11, 46
—— A, B lines, 55
—— F line, 60

Galaxy, 129
Galileo, 10
Gautier, 87
Gravitation, 41

INDEX

Hale, 76, 79
Halley, 105, 108
Hamlet, 3
Hansky, 69, 82, 112
Harvard College Observatory, 123
Helium, 54
Herschel, 11, 127, 131
Hirayama, 89
Horrebow, 83
Hydrogen, 55

Janssen, 56, 117

Kew, 98
Knowledge, a kaleidoscope, 7

Langley, S. P., 14, 69
Laplace, 110
Least action, 5, 8
Leonids, 102
Light, velocity of, 43
Lorentz, 64

Magnetic "character," 98
—— storms, 97
Maunder, E., 97
Mauritius, 86
Method, scientific, 1
Meton, cycle of, 107
Meudon, 74
Mont Blanc, 56
Moon, one of the planets, 31
Moon's motion, acceleration of, 109
Mount Wilson, 59

Newall, 116
Newcomb, 88, 108, 109, 117
Newton, dynamics, 5
—— hypotheses, 7
—— Queries, 8
Nicholson, 114

Oppolzer, 106
Oxygen, 57

Parallax, stellar, 130
Paschen, 57
Periodogram, 89
Perry, S. J., 117
Photosphere, 111
Planets, temperatures of, 22
Pouillet, 11
Principia, 7
Prominences, 87, 116

Queries, Newton's, 8

Radiation, pressure of, 22
Radium, 53
Rain bands, 55
Resonance, 50
Rigour, 8
Rowland, 47
Runge, 57

Saros, 105
Sabine, 12, 87
St John, 80
Saturn and Sun spots, 102
Schuster, 88, 89
Schwabe, 11, 83
Scientific method, 1
Smyth, C. P., 56
Solar apex, 127
—— constant, 18
—— motion, 11
Spectral type of stars, 122
Spectroheliograph, 72
Spectrum, 45
Spörer, 93
Stars, "fixed," 126
—— motions of, 126
—— spectral type, 122
—— spectral type and velocities, 130

INDEX

Stars, two streams, 129
Sun, age, 27
—— attraction on earth, 30
—— chemical elements, 51
—— circulation in, 81
—— contraction of, 25
—— density, 33
—— equator, 65
—— focus of the world, 4
—— gravity at, 34
—— heat how maintained, 28
—— internal mechanical state, 25, 35
—— mass, 33
—— oxygen in, 57
—— parallax, 36, 40
—— periodicities, 92
—— radiation, 18
—— radius, 33
—— rise of temperature from contraction, 27
—— rotation, 66

Sun, spots, 68
—— —— discovery, 10
—— temperature, 18

Taurus cluster, 128
Teneriffe, 56
Thales, 105
Thierkreis, 103
Three Bodies, 5
Turner, H. H., 102

Variable stars, 102

Wilson's spot theory, 11
Whitney, Mount, 14
Wolf, 83
Wollaston, 49

Young, 60, 96

Zodiac, 103